Timothy Attanucci

The Restorative Poetics of a Geological Age

Paradigms

Literature and the Human Sciences

Edited by
Rüdiger Campe · Paul Fleming

Volume 11

Timothy Attanucci

The Restorative Poetics of a Geological Age

Stifter, Viollet-le-Duc, and the Aesthetic Practices of Geohistoricism

DE GRUYTER

ISBN 978-3-11-068553-4
e-ISBN (PDF) 978-3-11-068947-1
e-ISBN (EPUB) 978-3-11-068951-8
ISSN 2195-2205

Library of Congress Control Number: 2020938895

Bibliografic information published by the Deutsche Nationalbibliothek
The Deutsche Nationalbibliothek lists this publication in the Deutsche Nationalbibliografie; detailed bibliografic data are available on the Internet at http://dnb.dnb.de.

Cover image: “A manual of the Mollusca”, Joseph Wilson Lowry, 1868 / Harvard University, Museum of Comparative Zoology, Ernst Mayr Library
Printing and binding: CPI books GmbH, Leck

www.degruyter.com

FSC www.fsc.org MIX Papier aus verantwortungsvollen Quellen FSC® C083411

Acknowledgements

First of all, I would like to thank the people whose vision and industry have made the publication of this book possible: Rüdiger Campe and Paul Fleming for conceiving the series *Paradigms* and accepting my study for publication; the anonymous peer reviewers, who generously gave their time to critique and improve my work; and the editors at de Gruyter, Lydia White, Myrto Aspioti, Stella Diedrich and Florian Ruppenstein, who have adeptly and amiably helped me throughout the publication process.

This book began as the third chapter of my dissertation, *Stories from Earth: Adalbert Stifter and the Poetics of Earth History*, which I submitted to Princeton University in 2012. Although the book represents a major transformation of that work, I would like to acknowledge here the institutional support I received from the German Academic Exchange Service (DAAD), the Whiting Foundation, the Einstein Foundation, and Princeton University. This generous support made my research possible at a crucial stage in my career. I would also like to thank the members of the Princeton German Department for their teaching and mentorship, in particular Stanley Corngold, who guided me through my first years of graduate study; Juliane Vogel, who first encouraged me to read Adalbert Stifter; and Thomas Levin, who shared enthusiasm, critique, and professional support in equally generous measures. I also thank Devin Fore, who, along with Thomas Levin, provided timely and essential insights as a reader of my dissertation. I owe special thanks to my two dissertation advisors, Joseph Vogl and Nikolaus Wegmann, both of whom not only provided me essential impulses, encouragement, and excellent advice as I wrote the dissertation, but have also been tireless in their mentorship and support in the years since. The other people who made my time in graduate school such a memorable and rewarding experience are too numerous to name here. In addition to my friends at Princeton and in the PhD-Net: *Das Wissen der Literatur* at the Humboldt University in Berlin, I would like to thank Megan Ewing, Alana King, and Cornelius Reiber, friends who not only made Princeton an exciting place to be a graduate student, but who have consistently lifted me up in the tough times that accompany the writing of a dissertation and the search for employment as a humanities PhD.

At the Johannes Gutenberg University in Mainz, I have had the joy of working with a whole new group of friendly and supportive colleagues, and in particular, with Christopher Busch, Matthias Emrich, and Loreen Dalski, who have all enriched this book with their insights over the last five years. I would like to give special thanks to Ulrich Breuer, who has generously advised me as I brought this project to fruition. His exemplary mentorship, moreover, has made me confident

https://doi.org/10.1515/9783110685534-001

that I will soon be able to thank him for his support of another text, my next book on melancholy and economics, which began as an article we co-authored for the *Handbuch Literatur und Ökonomie* (edited by Joseph Vogl and Burkhardt Wolf and published by De Gruyter in 2020). In addition to my coworkers in Mainz, a number of colleagues in Germany and the United States have provided invaluable support for my research in the form of lecture and conference invitations, including Anders Engberg-Pedersen, Christiane Frey, Michael Gamper, Arne Höcker, Frederic Ponten, Jana Schuster, Peter Schnyder, and Nathan Taylor.

I have saved the acknowledgments of my greatest debts for last: to my sister, Cara, who, as I write this, is risking her health during a pandemic to bring new lives safely into the world, but who has always found the right words of encouragement for her brother, who, on his best days, brings a new phrase onto the page; to my parents, John and Jane, who have themselves, in the midst of this crisis, turned their attention to the correction of this manuscript. Their editing help is only the tip of the iceberg – to explore its vast extent would require at least another book – but it suggests the gift they have provided their children: whenever I turn to them, they respond to the best of their ability, which includes a great capacity for love. Finally, I thank my wife Justine, who has graciously conversed with me about Stifter, Viollet-le-Duc, and geology since we first met. She also read the complete manuscript before I submitted it for review, and improved it immensely. More importantly, she helped convince me that, despite the flaws that remain and are my own, my desire to publish this study was not misplaced.

Contents

Introduction

The Human Age and the Geological Age

Since the start of this century, an academic debate among geologists has attracted a surprising amount of media and public attention. At issue is whether to rename the current geological epoch from the Holocene, which simply and indeterminately means "the most recent age," to the Anthropocene, or "the human age." The Nobel-prize winning Dutch atmospheric chemist Paul Crutzen has played a crucial role in popularizing the new term.[1] As he argued in *Nature* magazine in 2002: "Because of these anthropogenic emissions of carbon dioxide, global climate may depart significantly from natural behaviour for many millennia to come. It seems appropriate to assign the term 'Anthropocene' to the present, in many ways human-dominated, geological epoch."[2] While yet to be officially accepted as part of the official Geologic Time Scale set by the International Commission on Stratigraphy, many scientists, philosophers, science journalists, environmentalists and environmental humanists have eagerly adopted the concept. Historians have debated less whether the term is appropriate so much as when exactly the Anthropocene began. Was it two hundred years ago with the invention of the steam engine in 1784, as Crutzen proposed, or ten thousand years ago with the advent of agriculture, or even further back, when early humans began their upright march to planetary domination? While discussing the Anthropocene and its variants, philosophers have questioned not so much the impact of human activity on the global climate so much as the very meaning of categories "human," "earth," and "nature" in an age when such distinctions appear increasingly irrelevant.[3]

For someone unfamiliar with the term, its widespread success and impact, which at time reaches levels of alarm, might appear strange. After all, what is so remarkable about humans deciding to designate the age in which they live as the *human* age? Those versed in the tradition of critical theory might indeed

1 The concept was first introduced in a article co-written by Crutzen and Eugene F. Stoermer, "The Anthropocene," *IGBP Newsletter* 41 (2000): 17–18.

2 Paul J. Crutzen, "Geology of Mankind," *Nature* 415, no. 3 (2002): 23.

3 Donna Haraway has argued for a less comforting, less anthropo-centric term, the "Chthulucene," cf. *Staying with the Trouble: Making Kin in the Chthulucene* (Durham: Duke UP, 2016). Another critical variant is Jussi Parrika, *The Anthrobscene* (Minneapolis: U of Minnesota P, 2014). For an overview of the concept and a critical discussion of its relevance for the humanities, see Eva Horn and Hannes Bergthaller, *The Anthropocene: Key Issues for the Humanities* (London/New York: Routledge, 2020).

https://doi.org/10.1515/9783110685534-002

wonder whether talk of the Anthropocene is just another example of humanity's well-documented anthropocentrism.[4] Yet the clue to the uproar over the Anthropocene does not lie in the *anthropos* so much as in the *-cene*, and not in its meaning as "age," but in the fact that the suffix belongs to the academic jargon of geologists. To understand the Anthropocene and its cultural impact, in other words, it is essential to understand the meaning of geology.

I intend this book as a contribution to this endeavor. Its historical scope, the mid-nineteenth century, falls within the Anthropocene, and indeed, as Crutzen and Stoermer note, the effects of human activity on climate and landscape did not pass unobserved by the first generations of industrial modernity.[5] I have chosen to call this period "a geological age" rather than a human age, however, because I think that the significance of geology was much clearer to many in the nineteenth century than it is today. Indeed, in order to understand the drama and the pathos of the Anthropocene, including the magnitude of the decision before the International Commission of Stratigraphy, it is helpful to think back to and along with the nineteenth century, a time when the cultural fascination with geology arguably rivaled our interest in climate change today.

By "geological age," in any case, I mean to designate a period in which an author could call, as Honoré de Balzac did in his novel *La peau de chagrin* (*The Wild Ass's Skin*, 1831), the geologist and paleontologist Georges Cuvier "le plus grand poète de notre siècle" [the greatest poet of our age] for having "reconstruit des mondes avec des os blanchis" [reconstructed whole worlds out of bleached bones].[6] Balzac is hardly the only nineteenth century writer fascinated by geology.[7] In his novel *Der Nachsommer* (*Indian Summer*, 1857), which constitutes the

4 Cf. Hannes Bajohr, "Keine Quallen. Anthropozän und Negative Anthropologie," *Merkur* 73, no. 804 (2019): 63–74; reissued as the introduction to H. Bajohr (ed.), *Der Anthropos im Anthropozän: Die Wiederkehr des Menschen im Moment seiner vermeintlich endgültigen Verabschiedung* (Berlin/Boston: De Gruyter, 2020).

5 Crutzen and Stoermer, "The Anthropocene," 17.

6 Honoré de Balzac, *La Comédie Humaine*, vol. 10, ed. Pierre-Georges Castex (Paris: Gallimard, 1979), 75; *The Wild Ass's Skin*, ed. by Patrick Coleman, trans. by Helen Constantine (Oxford: Oxford UP, 2012), 19.

7 The research field of "geology and literature" is rapidly growing. In addition to the literature otherwise cited, see, in the British context, Noah Heringman's ground-breaking *Romantic Rocks, Aesthetic Geology* (Ithaca, NY: Cornell UP, 2004) and, with a focus on the mid-nineteenth century: Adelene Buckland, *Novel Science: Fiction and the Invention of Nineteenth-Century Geology* (Chicago: U Chicago P, 2013). Jörg Dünne, *Die katastrophische Feerie: Geschichte, Geologie und Spektakel in der modernen französischen Literatur* (Konstanz: Konstanz UP, 2017), traces the cultural impact of Cuvier's catastrophism on French literature. With a German and comparative focus: Peter Schnyder, ed., *Erdgeschichten. Literatur und Geologie im langen 19. Jahrhundert*

center of gravity of this study, the Austrian writer Adalbert Stifter found these words for a scientific field that in the preceding century and a half has probably elicited more than a few yawns among the uninitiated: "Wenn eine Geschichte des Nachdenkens und Forschens werth ist, so ist es die Geschichte der Erde, die ahnungsreichste, die reizendste, die es gibt" [If any history is worth pondering, worth investigating, it is the history of the Earth, the most promising, the most stimulating history there is].[8] The novel's narrator is a geologist, so his fascination with the history of the earth is to be expected, but these superlatives suggest something more. Earth history is not only a "stimulating" field – the word *reizend* suggests sensual pleasure – "worth investigating," but also a "promising" history – a better translation of *ahnungsreich* might be "portentous" – "worth pondering" (*Nachdenken*).

The portent of earth history lies in its relation to human history, for the earth has "eine Geschichte, in welcher die der Menschen nur ein Einschiebsel ist, und wer weiß es, welch ein kleines, da sie von anderen Geschichten vielleicht höherer Wesen abgelöset werden kann" [a history where man is only an interpolation, who knows how small a one, and can be superseded by other histories of perhaps higher beings].[9] This history is incommensurably longer and greater than the history of humanity, which shrinks, in comparison, to a mere "interpolation" (*Einschiebsel*). The world *before* humans also implies a world *after* humans.

This marginalization of humanity to a razor's edge of earth history, so eloquently expressed in Stifter's literary works, helps explain why the Anthropocene is, in turn, such a momentous and scandalous development. If we are just visitors on this planet, with what right do we determine its geological fate? For writers in the nineteenth century, the newly discovered possibility of a "plot without man" was itself the scandal.[10] Thanks to the paleontologist Steven Jay Gould and Sigmund Freud, to whom Gould turns to explain the impact of geology on our collective self-understanding, we have a convenient name for this

(Würzburg: Königshausen & Neumann, 2020) and Jason Groves, *The Geological Unconscious: German Literature and the Mineral Imaginary* (New York: Fordham UP, 2020), which includes a chapter on Stifter.

8 Adalbert Stifter, *Werke und Briefe: Historisch-Kritische Gesamtausgabe*, vol. 4.2, edited by Alfred Doppler, Wolfgang Frühwald, and Hermann Kunisch (Stuttgart: Kohlhammer, 1978), 32. Hereafter with the abbreviation HKG; Stifter, *Indian Summer*, trans. Wendell Frye (Bern: Peter Lang, 2009), 445. All further citations of this (and other published translations cited), will be marked with page number in brackets after the citation of the original.

9 Ibid.

10 Gillian Beer, *Darwin's Plots: Evolutionary Narrative in Darwin, George Eliot and Nineteenth-Century Fiction*, revised ed. (Cambridge: Cambridge UP, 2000), 17.

scandalous discovery: the geological "outrage upon [our] naïve self-love."[11] Like Copernicanism, Darwinism and psychoanalysis, which constitute Freud's list of the most significant "narzisstische Kränkungen," the discovery of so-called "deep time" represents a massive disruption to anthropocentrism.[12] As a younger American contemporary of Stifter, Mark Twain, elegantly illustrates, the brute fact of a hundred million years of prehistory challenges any notion that the earth exists *for* us:

> Man has been here 32,000 years. That it took a hundred million years to prepare the world for him is proof that that is what it was done for. I suppose it is, I dunno. If the Eiffel Tower were now representing the world's age, the skin of paint on the pinnacle-knob at its summit would represent man's share of that age; and anybody would perceive that that skin was what the tower was built for. I reckon they would, I dunno.[13]

The numbers, of course, have changed since 1903, but the enormous disproportion of the time *before* humanity and the time *of* humanity, especially when compared with the Biblical creation ratio of five days to one, remains a central component of our current worldview. As Twain's humorous remark makes clear, moreover, the scandal of deep time lies not so much in the claim that humanity is small or insignificant in some existential or metaphysical way, but rather that we literally do not *count:* on a uniform, material plane of being, our collective existence amounts, quantitatively, to an infinitesimal fraction of the whole, in other words: zero.

The German philosopher Friedrich Nietzsche, who counts among Adalbert Stifter's most famous and most adoring readers, was also fascinated by the at times dizzying and chilling notion of "deep time."[14] It is tempting to speculate that Nietzsche, reading Stifter's *Indian Summer*, which the philosopher claims belongs to only a handful of modern German books which "deserve to be read

---

11 Stephen Jay Gould, *Time's Arrow, Time's Cycle: Myth and Metaphor in the Discovery of Geological Time* (Cambridge, MA: Harvard UP, 1988), 2.

12 Sigmund Freud: "Eine Schwierigkeit der Psychoanalyse." In: *Gesammelte Werke*, Vol. 12 (Frankfurt a.M.: Fischer 2005), 7. The phrase "deep time" was popularized by John McPhee, *Basin and Range* (New York: Farrar, Straus, Giroux, 1981).

13 Cited in Gould, *Time's Arrow*, 2.

14 A prominent example in German literature is Annette von Droste-Hülshoff's poem *Die Mergelgrube* (1842). For a good overview of such examples, see Georg Braungart, "Apokalypse in der Urzeit. Die Entdeckung der Tiefenzeit in der Geologie um 1800 und ihre literarischen Nachbeben," in *Zeit-Zeitenwechsel-Endzeit*, ed. G. Leinsle and Jochen Mecke (Regensburg: Universitätsverlag 2000), 107–120.

again and again,"[15] was delighted to find confirmation of an idea that would become crucial to his philosophical thinking. In any case, Nietzsche's commentary on "deep time" is one of the more memorable and consequential in the history of German literature. The following "fable," as Nietzsche calls it, would be unthinkable without the discoveries of modern geology:

> In irgend einem abgelegenen Winkel des in zahllosen Sonnensystemen flimmernd ausgegossenen Weltalls gab es einmal ein Gestirn, auf dem kluge Thiere das Erkennen erfanden. Es war die hochmüthigste und *verlogenste* Minute der 'Weltgeschichte': aber doch nur eine Minute. Nach wenigen Athemzügen der Natur erstarrte das Gestirn, und die klugen Thiere mussten sterben.
>
> In some remote corner of the universe, poured out and glittering in innumerable solar systems, there once was a star on which clever animals invented knowledge. That was the haughtiest and most mendacious minute of 'world history' – yet only a minute. After nature had drawn a few breaths the star grew cold, and the clever animals had to die.[16]

This pointed formulation of the geological outrage suggests the wide range of its potential implications. If our animal existence has no privileged position in some chain of being, then what of our intellectual existence? The question leads Nietzsche to the elaboration of his radical perspectivism: even a mosquito, after all, probably sees itself as the center of existence. The image of the inevitable death of the "clever animals," moreover, recalls not only Stifter's question quoted above – who knows if other, higher beings might one day replace us? – but also prefigures Nietzsche's *Übermensch*, an aspirational formulation of the post-human.

---

15 Beyond Goethe's works, including his *Conversations with Eckermann* – "dem besten deutschen Buche, das es giebt" [the best German book there is] – Nietzsche asks: "was bleibt eigentlich von der deutschen Prosa-Litteratur übrig, das es verdiente, wieder und wieder gelesen zu werden?" [What works are left of German prose literature that deserve to be read again and again?]. He answers with a remarkably short list: "Lichtenberg's Aphorismen, das erste Buch von Jung-Stilling's Lebensgeschichte, Adalbert Stifter's Nachsommer und Gottfried Keller's Leute von Seldwyla, – und damit wird es einstweilen am Ende sein" [Lichtenberg's aphorisms, the first book of Jung-Stilling's autobiography, Adalbert Stifter's *Indian Summer* and Gottfried Keller's *The People of Seldwyla*, – and that would be about it for now]. Nietzsche, "Menschliches Allzumenschliches II: § WS – 109" [1879] in: *Digital Critical Edition of the Complete Works and Letters*, based on the critical text by G. Colli and M. Montinari (Berlin/New York: De Gruyter 1967–), edited by Paolo D'Iorio, http://www.nietzschesource.org/#eKGWB/WS-109.

16 Friedrich Nietzsche, "Über Wahrheit und Lüge im außermoralischen Sinne: § 1," http://www.nietzschesource.org/#eKGWB/WL-1. "On Truth and Lying in an Extra-Moral Sense," in *The Portable Nietzsche*, trans. Walter Kaufmann (New York: Viking Penguin, 1977), 42.

The topos of deep time as an outrage in response to our human narcissism is still a central component of the modern, scientific worldview. The discourses of science pedagogy and popularization have not tired of images which, like Stifter's "interpolation," Twain's "skin of paint" on the Eiffel Tower, or Nietzsche's mosquito, "the flying center of this world,"[17] aim to illustrate the scale of the history of the universe. In his 2014 revival of Carl Sagan's TV-series *Cosmos*, for example, Neil deGrasse Tyson returned to the metaphor of the cosmic calendar to hit home this point. If one imagines the history of the universal on the scale of a calendar year, with the Big Bang occurring at midnight on January 1, so Tyson explains, then the first human cave paintings do not appear until the very last minute of that year. The necessity of such images clearly stems from our inability to imagine the reality of 13.82 billion years of cosmic history.

The concept of the Anthropocene adds an important twist to this modern story. As hard as it may have been to accept that we are a mere afterthought of geological time, we now have to come to terms with the fact that our actions, via climate change, are having an impact on the planet that is *geologically significant*. As short as it is on the vast scale of earth history, human history has still inaugurated, with the Anthropocene, a new geological epoch. The pathos of the Anthropocene resides in a paradox that only a geologist can fully appreciate: on a scale where we are supposed to be insignificant, a minimal factor that can be ignored, we are in fact having a real impact.

The reality of climate change forces us to reconsider our relationship to nature.[18] We can no longer see the planet as part of a nature that remains outside of us because it is and will forever be greater than we are. The message of the new ecologies is that the entire planet is our *oikos*, our house, in an emphatic sense. It is not only what surrounds or underlies our existence, but it is a milieu proximate enough that our life changes it. Just as a mosquito is bound to its pond, or the bat to the mosquito population, so too are we bound to the limits of our biosphere.

This reevaluation of the planetary effects of human industry suggests that a certain narrative of the history of geology is in need of revision. As an "outrage to our naïve self-love," geological deep time indeed appears analogous to the Copernican and Darwinian revolutions. As challenges to the anthropocentric world-

---

17 Ibid.

18 Accordingly, the field of *ecocriticism* is booming. For orientation, one can turn to any number of introductions and surveys such as Greg Garrard, *The Oxford Handbook of Ecocriticism* (New York: Oxford UP, 2014); Benjamin Bühler, *Ecocriticism: Grundlagen – Theorien – Interpretationen* (Stuttgart: Metzler, 2016); and Gabriele Dürbeck and Urte Stobbe, eds., *Ecocriticism: Eine Einführung* (Cologne: Böhlau, 2015).

view, these insights represent important steps on the path to secular Enlightenment. As we know, if only from Theodor W. Adorno and Max Horkheimer's classic study, Enlightenment has a curious dialectic. Just as Copernican astronomy ultimately leads to space flight and Darwinian biology to genetic engineering, so too does the digging of the nineteenth century, the unearthing of geological evidence, fossils and fossil fuels at once, lead to the paradox of the Anthropocene. In this respect, all these "outrages" do indeed seem comparable to Freudian psychoanalysis, which not only holds that the "I" is not "master in his own house" – the phrase most people will remember from the essay "A Difficulty of Psychoanalysis" – but also that the other, whatever "it" is, is also inside of "me." The outrage of climate change is thus not simply that we have caused environmental damage, but that we have set in motion a process of geological change that may now be beyond our control.

In its relationship to the present, and when considered on a global scale, the aesthetic attitude that I argue emerged in a prior geological age, namely "restorative poetics," has similarly wide implications. Indeed, now that we have recognized our overwhelmingly destructive relationship to the earth and its ecosystems, there is perhaps no other imaginable alternative than a new approach that is, broadly speaking, "restorative." At least, it is now commonplace to speak of the restoration of landscapes and ecosystems, and larger-scale forms of geoengineering are on the horizon.

Two Avatars of Geohistoricism: Stifter and Viollet-le-Duc

While this broader, contemporary context provides the backdrop and the motivation of the following study, its scope is considerably more modest. Historically speaking, the geological age at issue here extends from the 1830s to the 1880s, a "short" nineteenth century, so to speak, that encompasses the active working years of its two main protagonists, Adalbert Stifter (1805–1868) and Eugène-Emmanuel Viollet-le-Duc (1814–1879). The mid-nineteenth century was not only the heyday of widespread public fascination with geology,[19] but also of the mentalities and discourses collectively known as historicism. I call the intersection of the fascination with earth history, antiquities, and historical relics of all kinds "geohistoricism." Although geohistoricism includes the root word "historicism" or *Historismus*, I define this much-discussed concept as broadly

19 Cf. Ralph O'Connor, *The Earth on Show: Fossils and the Poetics of Popular Science, 1802–1856* (Chicago: U Chicago P, 2007).

as possible.[20] If historicism can be considered "a mode of historical thought developed primarily in the nineteenth century," then I do not see the indefinite article as an invitation to quarrel about definitions of *the* (authentic or ideal) historicism, so much as an opening to a plurality of historical forms.[21] Geohistoricism certainly falls within the breadth of a general "historicism [that] has to do with explanation or evaluation by means of history and with the belief that historical knowledge is in some sense distinctively important in human affairs," and extends the relevance of "historical knowledge" to earth history.[22] It also participates in, or contributes to, the sense of relativism that characterizes many forms of historicism.[23] Above all, however, I understand the neologism not so much as a challenge to many concepts of historicism in fields such as philosophy, history, economics, law, art and literary history, but as an invitation to scholars in these fields to consider the profound meaning of earth history and its geological imagination for the historical imagination of the nineteenth century on the whole.[24]

My interest lies, moreover, not so much in a set of ideas or concepts collectively called geohistoricism as its *aesthetic practices*. In this sense, geohistoricism may be called a discourse, but I am interested not only in the rules of what can

---

20 My approach is indebted to Chenxi Tang's critical revision of "the standard narrative of the rise of historism," which aims to reintegrate "a vital dimension of modern historical thought, namely, the conception of history as intertwined with and shaped by geographic conditions." Nevertheless, I use the term "geohistory" more narrowly, that is, to indicate a *geological*, rather than a broadly "geographic," mode of history. Cf. Chenxi Tang, *The Geographic Imagination of Modernity: Geography, Literature, and Philosophy in German Romanticism* (Stanford UP, 2008), 203; for the entire discussion of historism/historicism, 202–239.

21 "Eine vornehmlich im 19. Jahrhundert ausgebildete Weise des historischen Denkens," writes Michael Schlott, "Historismus," ed. Harald Fricke et al., *Reallexikon der Deutschen Literaturwissenschaft,* (Berlin: De Gruyter, 1997), 58. For a history of the concept with detailed discussion of the controversies involved, see Annette Wittkau-Horgby, *Historismus: zur Geschichte des Begriffs und des Problems* (Göttingen: Vandenhoeck und Ruprecht, 1994) and beyond their position outside the philosophical and historical faculties, Stifter and Viollet-le-Duc are also historically situated, following Wittkau-Horgby, in the period of the emergence of historicism, that is, after the early Romantics coined the term, but before professional scholars institutionalized it as a method.

22 Dwight E. Lee and Robert N. Beck, "The Meaning of 'Historicism,'" *The American Historical Review* 59, no. 3 (1954): 569.

23 Ibid. 573–574.

24 There is, for example, considerable overlap between Stifter and Viollet-le-Duc's painterly engagements with history and the 'visual culture' of academic historicism examined by Kathrin Maurer, *Visualizing the Past: The Power of the Image in German Historicism* (Berlin/Boston: de Gruyter, 2013).

be said and thereby known, but also in the poetic forms that generate their own specific knowledge.[25] In his various roles as writer and landscape painter, as art conservator and school overseer, and as amateur natural scientist, Adalbert Stifter exemplifies various key aesthetic practices of geohistoricism. In his first novella, *The Condor* (1840), Stifter establishes an interest in the aesthetics of natural science, and particularly the earth sciences, that he will pursue throughout his oeuvre, from the novellas collected in his *Studien* (in several volumes, 1844–1850) and *Bunte Steine* (*Many-Colored Stones*, 1853), to his *Bildungsroman* about a geologist (*Indian Summer*, 1857), and his grand attempt at an authentic restoration of the Middle Ages in the historical novel *Witiko* (1867). With a title that references the work of the period's most famous German naturalist and geologist, Alexander von Humboldt, *The Condor* questions the role of landscape painting in physical geography, as well as the (in)visibility of meteorological phenomena. Stifter's best-known work, *Many-Colored Stones*, employs the geological names of stones, such as *Granite*, *Limestone*, *Rock Crystal*, and *Tourmaline*, as titles for his stories, many of which he had previously published under different names in journals. Often, this symbolic use of geological terms is superficial, and Stifter's narrators comment on the fact, in *Tourmaline*, for example: "Der Turmalin ist dunkel, und was da erzählt wird, ist sehr dunkel" [Tourmaline is dark, and this story here is very dark].[26] In his introduction to the work, moreover, Stifter comments on having collected stones as a child, and often including a weathered piece of glass in his collection, merely for its beauty. In examining the geologics of Stifter's narratives, one must have as much of an eye for hidden structures and aesthetic forms as to the more obviously "geological" signs. The analogy between the natural and the human world, for example, expressed in the concept of the "gentle law" in Stifter's "Preface" to *Many-Colored Stones*, is prevalent throughout author's literary oeuvre and parallels the embedding of human history within earth history as an "interpolation" in *Indian Summer*.[27]

25 In this sense, I am interested not simply in a Foucauldian "archeology of knowledge," but in the "poetics of knowledge," cf. Joseph Vogl, "Poetologie des Wissens," in *Einführung in die Kulturwissenschaft*, ed. Harun Maye and Leander Scholz (Munich: Fink, 2011), 49–71.

26 HKG 2.2, 135.

27 Eric Downing, for example, turns to this analogy to demonstrate how the granite stone serves as a "threshold" between the present and the past, natural and cultural *Urgeschichte*, cf. E. Downing, "With Wandering Steps and Slow: Schwellenkunst in Adalbert Stifter's Granit," *DVjs* 94 (2020): 75. On geology in *Many-Colored Stones*, see also Jason Groves, "Erratische Blöcke: Stifters 'Bunte Steine,'" in *Sema: Wendepunkte der Philologie*, ed. Joachim Harst and Kristina Mendicino (Würzburg: Königshausen & Neumann, 2013), 255–269, and for a discussion of the meteorological underpinnings of the "gentle law," see my article, "'The Gentle Law' of Large Numbers: Stifter's Urban Meteorology," *Monatshefte* 112, no. 1 (2020): 1–19.

This study thus argues that the intersection of nature and culture, and in particular, of the discovery of earth history and desire for history in Stifter's works both exemplify and reflect upon the aesthetic culture of geohistoricism. The focal points of Stifter's geohistoricism are the works studied here, the novella *Die Narrenburg* (*Castle Crazy*, 1843), part of a convolute of texts associated with *My Great-Grandfather's Notebook*, as well as *Indian Summer* and *Witiko*. These three projects occupied Stifter throughout his writing career, and each is related to Stifter's restorative poetics. If *Castle Crazy* and the *Notebook* represent an attempt to reconstruct family history in the service of storytelling, *Indian Summer* reflects on the conditions of geohistoricist culture, its uses and values. As a historical novel, in turn, *Witiko* is applied geohistoricism.

In the 1850s, as Stifter wrote *Indian Summer* and *Witiko*, he was also officially named a *Conservator* of historical buildings and artworks by the Austrian government. The meeting between a geologist and an art conservator described in *Indian Summer* is thus broadly autobiographical. Although Stifter's interests in geology and in art restoration have both been explored as individual themes, there has not been, to my knowledge, any recognition of the deep connection of the two.[28] In arguing that what others have called his "aesthetic" or "museum" historicism is fundamentally geohistoricism,[29] this book aims, as a contribution to research on Stifter, to restore the coherency to his poetic projects, which, as such, represent his deep understanding of the foundations of nineteenth century historicist culture.

Stifter was not alone in connecting historicism and cultural conservation with the practices and insights of geology. Through a comparison of Stifter's literary descriptions with the aesthetic theory and practice of the French architect Viollet-le-Duc, it becomes possible to recognize the breadth of the discursive relationships between geology, historicism, and practices of reconstruction in the nineteenth century. Like Stifter, Viollet-le-Duc brought an interest in geology and

---

28 Because of its thematic focus, this study avoids a systematic review of literature on Stifter in favor of punctual citation. At this point, however, it is worth mentioning the articles on Stifter and geology by Peter Schnyder and Sabine Schneider, as well as the essay on restoration by Saskia Haag in Michael Gamper and Karl Wagner, eds., *Figuren der Übertragung: Adalbert Stifter und das Wissen seiner Zeit* (Zürich: Chronos, 2009). This volume, the first to apply the methodology of *Wissensgeschichte* [history of knowledge] to Stifter, lays out a discursive space in which the connection between geology and restoration becomes more easily visible. For an overview of current research, readers can consult the excellent *Handbuch* edited by Christian Begemann and Davide Guiriato, *Stifter-Handbuch: Leben – Werk – Wirkung* (Stuttgart: Metzler, 2017).
29 Cf. Heinz Schlaffer and Hannelore Schlaffer, *Studien zum ästhetischen Historismus* (Frankfurt a. M.: Suhrkamp, 1975) and Katharina Grätz, *Musealer Historismus: die Gegenwart des Vergangenen bei Stifter, Keller und Raabe* (Heidelberg: Winter, 2006).

landscape painting to his primary professional endeavors as an architect, conservator, and writer. Of course, Viollet-le-Duc's stature and impact on architectural restoration in France far exceeds the rather marginal practical role played by Stifter in Upper Austria. From the early 1840s onwards, the rising star of French architecture was entrusted with some of the nation's most important restoration projects, including the Sainte Chapelle, the Notre-Dame cathedral in Paris, the Basilica in Saint Denis, and the walls of Carcassonne. Perhaps as importantly, his monumental, ten-volume *Dictionnaire raisonné de l'architecture française du XIe au XVIe siècle* (*Dictionary of French Architecture from the Eleventh to the Sixteenth Century*, 1854–1868) not only set a new standard for historical knowledge of medieval architecture, but also inspired, in its visual style, an entire generation of modern historicist building in France. Bolstered by a six-volume *Dictionnaire raisonné du mobilier français de l'époque carolingienne à la Renaissance* (*Dictionary of French Furniture from the Carolingian Period to the Renaissance*, 1858–1870), the theoretical *Entretiens sur l'architecture* (*Interviews on Architecture*, vol. 1, 1863; vol. 2, 1872) and a series of popular, short *Histoires* (1873–1879) on topics such a house, a fortress, or the education of a *dessinateur* (draughtsman, designer),[30] Viollet-le-Duc's publications assured him a reputation in France that rivaled Gottfried Semper in Germany and John Ruskin in Britain and lasted until the broad cultural turn against historicism at the turn of the twentieth century.

While Semper's and Ruskin's writings provide further evidence for the varied connections between geology and historicist architecture, Viollet-le-Duc's vivid "scientific imagination" is particularly illuminating as regards the deeper discursive framework that informs both geological and historical reconstruction practices.[31] Indeed, he explicitly credits the geologist Georges Cuvier alone for the development of "une attitude inusitée" [an entirely new attitude] towards history that distinguishes the nineteenth century from all preceding periods.[32] For Viollet-le-Duc, not only does his own method – a comparative "anatomy" and evolutionary theory of architectural forms – and practice of historical restoration

30 The *Histoires* were published by Pierre-Jules Hetzel, in the same series as the science-fiction stories of Jules Verne.

31 I borrow the phrase from Laurent Baridon, *L'imaginaire scientifique de Viollet-Le-Duc* (Paris: Harmattan, 1996), who also discusses Viollet-le-Duc's relationship to Semper. For a comparison of Viollet-le-Duc and Ruskin with respect to geology, see Jan von Brevern, *Blicke von Nirgendwo: Geologie in Bildern bei Ruskin, Viollet-le-Duc und Civiale* (Munich: Wilhelm Fink, 2012).

32 Eugène-Emmanuel Viollet-le-Duc, "Restauration," *Dictionnaire raisonné de l'architecture française du XIe au XVIe siècle*, vol. 8 (Paris: Morel, 1858), 15. Viollet-le-Duc, *On Restoration*, trans. Charles Wethered (London: Sampson Low, Marston Low, and Searle, 1875), 13.

owe a great debt to Cuvier, so do a whole series of historical sciences, from archeology to the history of art and philology, to ethnology and even the history of science. What all these fields have in common, in addition to the historical "attitude" which seeks to analyze the past and reconstruct its development "step by step," is a new focus on the study of the materials or "things" of history. Viollet-le-Duc's restorations can thus be seen as a kind of *Leitmedium* for nineteenth century historicism. In reconstructed chapels and palace chambers, one rediscovers the past *in vivo*, as a historical "milieu," a world unto itself, a functional context. Following Viollet-le-Duc's suggestion, which closely matches Stifter's alignment of geology with art history in his novel *Indian Summer*, this study thus observes the historical imagination of the nineteenth century from the perspective of the history of knowledge.

Before turning to a preview of the chapters to come, and the deeper explorations of the constellation of geohistorical poetics that they provide, it is important to establish how, from this knowledge-historical perspective, the 1830s present a caesura. This delineation of the historical beginning of geohistoricism is also pressing because, in the minds of historians of the nineteenth century, the date 1830 already marks the date of political revolution, at least in France. What is more, the establishment of the more liberal July Monarchy in 1830 ended what historians call the Bourbon Restoration. In terms of personal biography, it is hardly a coincidence that a figure such as Viollet-le-Duc, who grew up during this *Restauration*, would later devote himself to architectural work that he would choose to call "restauration."[33] From a political perspective, it is also evident that Viollet-le-Duc's career, and indeed Stifter's work as a *Conservator* would have looked very different, at least in practice, without the nationalist cultural politics that drove the development of state institutions for cultural and historical conservation.[34]

At the same time, the history of knowledge, and in particular of geological knowledge, sheds a rather different light on the history of conservation and re-

33 As befits the psychologizing approach of biographies, the relationship between restoration and personal and national "mourning" is explored by Martin Bressani, *Architecture and the Historical Imagination: Eugène-Emmanuel Viollet-Le-Duc, 1814-1879* (Farnham UK: Ashgate, 2014), 3–45.

34 Stifter was fairly well connected among Viennese political and cultural elites in the 1830s. He briefly tutored the son of Metternich himself, and was a guest in the famous salon of Caroline Pichler. It was his connection to the historian and numismatist Joseph von Arneth and his wife Antonie (Adamberger), an actress also friendly with Pichler, that led to his official post as *Conservator*, see Wilfried Lipp, "Adalbert Stifter als 'Conservator' (1853–1865)," in *Sanfte Sensationen Stifter 2005: Beiträge zum 200. Geburtstag Adalbert Stifters*, ed. Johann Lachinger et al. (Linz: StifterHaus, 2005), 185–203.

storation, and indeed to the nineteenth century's broader historical imagination. Geology is hardly apolitical, as the famous debate between Vulcanists and Neptunists circa 1800 demonstrates. Vulcanists, who believed in a fiery origin of the earth, often held revolutionary political attitudes, while Neptunists, who imagined that the earth's surface crystallized from a primeval ocean, tended to prefer slower political reform as well.[35] Another distinction established by political readings of nineteenth-century geology sees in Cuvier's so-called catastrophism the embodiment of the spirit of the French Revolution, and in James Hutton's gradualism or Charles Lyell's uniformitarianism evidence of a more conservative British reaction. The history of knowledge offers a way of grouping these figures differently. Instead of two political camps, it sees two distinct debates, each of which might resonate with the political debates of its time.[36]

For the history of geology, the publication of the first volume of Lyell's *Principles of Geology* in 1830 is thus not a milestone because it introduces a conservative response to Cuvier's catastrophism. Instead, Lyell's masterwork represents the culmination of decades of work by geologists that established geology as a genuinely *historical* science. For Martin Rudwick, whose monumental two-book history of geology, *Bursting the Limits of Time* (2005) and *Worlds Before Adam* (2008), serves as a guide for the following, Georges Cuvier played a starring role in the transition from the paradigm of geology at the end of the eighteenth century, which Rudwick calls *geotheory* and is represented by scientists such as Georges-Louis Leclerc, Comte de Buffon and Abraham Gottlob Werner, to the *geohistory* practiced by Cuvier, Humboldt, and Lyell.[37]

This transition is also apparent in works of literature from the eighteenth and nineteenth century. Students of Werner at the Freiberg Mining Academy (*Bergakademie*) such as Friedrich von Hardenberg, known as Novalis, for example, brought their teacher's geotheoretical or geognostic ideas into central texts

35 On the lasting impact of this scientific-political debate, see Werner Michler, "Vulkanische Idyllen: Die Fortschreibung der Revolution mit den Mitteln der Naturwissenschaft bei Moritz Hartmann und Adalbert Stifter," in *Bewegung im Reich der Immobilität – Revolutionen in der Habsburgermonarchie 1848-1849; literarisch-publizistische Auseinandersetzungen*, ed. Hubert Lengauer and Primus-Heinz Kucher (Vienna: Böhlau, 2001), 472–495. For a social history of geology, see Roy Porter, *The Making of Geology: Earth Science in Britain, 1660–1815* (Cambridge, UK: Cambridge UP, 1977).

36 As Martin Rudwick notes in his "Introduction" to the *Principles*, Lyell opposes catastrophism not for political reasons, but because it "inhibits discovery," Charles Lyell, *Principles of Geology, Volume 1* (Chicago: U Chicago P, 1990), xxxviii.

37 Martin Rudwick, *Bursting the Limits of Time: The Reconstruction of Geohistory in the Age of Revolution* (Chicago: U Chicago P, 2005), esp. 5–8 and 133–138; and Rudwick, *Worlds Before Adam: The Reconstruction of Geohistory in the Age of Reform* (Chicago: U Chicago P, 2008).

of early German Romanticism.[38] From the perspective of German literary and cultural history, of course, it is also common to refer to the "Age of Goethe" or *Goethezeit* (born in 1749, the author died in 1832). This period also corresponds with the major social transformations "circa 1800" that Reinhart Koselleck calls the *Sattelzeit* [Saddle Period] and with the transformation of knowledge discourses that Michel Foucault describes in *The Order of Things*.[39] In the case of the pre-history of geohistoricism, the polymath Goethe's writings on geology can serve as an introduction to this transition, which itself remains outside the scope of this study. The difference between Goethe's and Stifter's approach to geology, however, underscores the importance of the 1830s as a caesura in literary history, which conventional literary histories would describe as the shift from romanticism to realism. In this regard, the aesthetic practices of geohistoricism present a new way to understand (German) realism's obsessive interest not only in antiquities, but indeed "things" of all kinds. Indeed, in light of geohistoricism as a slice in the history of knowledge, one might even speak of an "Age of Stifter" from ca. 1830 to 1870.[40]

The Historicization of Nature: Antiquarianism, Restoration, and Historicism in Geology

During the Age of Goethe, the study of the earth changed dramatically. While Buffon's *Époques de la Nature* (1778) deserves a place in the history of science for promoting the idea of an ancient earth, the French naturalist's foray into the geological past still has at least one foot firmly planted in the "classical age" of eighteenth-century natural history and its static tableaus of knowledge.[41] Buffon's approach also belongs to what Rudwick calls geotheory as opposed to

38 On the central role played by the geologist Werner in German Romantic circles, see Michaela Haberkorn, *Naturhistoriker und Zeitenseher: Geologie und Poesie um 1800. Der Kreis um Abraham Gottlob Werner (Goethe, A.v. Humboldt, Novalis, Steffens, G.H. Schubert)* (Frankfurt a.M.: Peter Lang, 2004).

39 Cf. Reinhart Koselleck, "Introduction and Prefaces to the Geschichtliche Grundbegriffe," trans. Michaela Richter, *Contributions to the History of Concepts* 6, no. 1 (2011): 1–37. Michel Foucault, *The Order of Things: An Archaeology of the Human Sciences* (London: Routledge, 2005).

40 The phrase "Age of Stifter" or "Stifter-Zeit" was coined (playfully) by Michael Gamper and Karl Wagner, eds., *Figuren der Übertragung*, 8. The history of knowledge cuts up the past differently than the hagiographic author philology behind the 'Age of Goethe.'

41 In addition to Foucault, *The Order of Things*, see also Wolf Lepenies, *Das Ende der Naturgeschichte: Wandel kultureller Selbstverständlichkeiten in den Wissenschaften des 18. und 19. Jahrhunderts* (Munich: C. Hanser, 1976).

geohistory. From Thomas Burnett's *Sacred Theory of the Earth* (1681/1690) to the debate between Vulcanists and Neptunists circa 1800, the goal of geology was to explain the earth's origins, its primordial formation. Such theory was often speculative in character, but geohistory is not merely more empirical than earlier theorizing. The defining feature of historical geology as practiced by Cuvier or Lyell is their new conceptualization of their object of study as *historical*. They do not study the earth's formation as a singular process, be it cooling or crystallization, but rather the earth's history as a series of contingent events and distinct epochs. For Lyell, modern geology thus begins with James Hutton's declaration "that geology was in no ways concerned 'with questions as to the origin of things.'"[42] In this sense the unity of geohistory does not appear on the level of object of study – is the *earth as a whole* primarily volcanic or did it emerge from the primordial ocean? – but rather, on the level of method. In the theory of the earth, it was possible, indeed expedient, to assume that earth's past may have been fundamentally different from its present, both in terms of surface effects and underlying causes. By contrast, Lyell's famous motto, "the present is the key to the past," depends on the rigorous separation between physical theory, which explains what *can* happen, and historical research, which explains what *did* happen.

Both Cuvier and Lyell saw their work on earth history or geological prehistory as comparable to the work of other historians. Considering the full title and first pages of Cuvier's *Recherches sur les ossements fossiles des quadrupèdes* (*Researches on the Fossil Bones of Quadrupeds*, 1812), one can easily see how Viollet-le-Duc could recognize Cuvier as the founding father of restoration. The subtitle reads: "Ou l'on rétablit les caractères de plusieurs espèces d'animaux que les révolutions du globe paroissent avoir détruites" [In which we reestablish the characteristics of several species of animal that the revolutions of the earth appear to have destroyed].[43] The goal of Cuvier's paleontology is to reconstruct what geological revolutions have destroyed. And as the first paragraph of the *Discours Préliminaire* makes clear, no number of words with the prefix *re-* is too great to express Cuvier's interest in recovering what appears to have been lost or destroyed:

> J'essaie de parcourir une route où l'on n'a encore hasardé que quelques pas, et de faire connoître un genre de monumens presque toujours négligé, quoique indispensable pour l'histoire du globe.

42 Lyell, *Principles of Geology*, 4.

43 Georges Cuvier, *Recherches sur les ossements fossiles des quadrupèdes* (Paris: Deterville, 1812), title page.

> Antiquaire d'une espèce nouvelle, il m'a fallu apprendre à déchiffrer et à *restaurer* ces monumens, à *reconnoître* et à *rapprocher* dans leur ordre primitif les fragmens épars et mutilés dont ils se composent; à *reconstruire* les êtres antiques auxquels ces fragmens appartenoient; à les *reproduire* avec leur proportions et leur caractères; à les comparer enfin à ceux qui vivent aujourd'hui à la surface du globe: art presque inconnu, et qui supposoit une science à peine effleurée auparavant, celle des lois qui président aux coexistences des formes des diverses parties dans les êtres organisés.
>
> I attempt to trace a path down which, until now, one has only dared to take a few steps, and to make known a genre of monuments that has been almost entirely neglected, despite how indispensable it is for the history of the planet.
>
> As the antiquarian of a new species, I needed to learn how to decipher and to *restore* these monuments, to *recognize* and *return* to their primitive order dispersed and mutilated fragments of which they are composed; to *reconstruct* the antique beings to which these fragments belong; to *reproduce* them with their proportions and characteristics; to compare them, finally, with those who live today on the surface of the planet: a technique almost completely unknown, and which supposes a science that until now has barely been touched upon, that of the laws that preside over the coexistence of forms of the different parts of organic beings.[44]

In this passage, Cuvier evidently sees antiquarianism and the restoration of historical monuments as the model for his work in reconstructing fossil bones into complete skeletons. Although Viollet-le-Duc refers to Cuvier as the source of his century's historical "attitude," the relationship between geology and historicism is thus not one of direct influence or even precedence. As Noah Heringman has recently argued, several discourses circa 1800, among them antiquarian collecting, classical philology, and natural history, all employ the word "antiquity" to designate their object of study.[45] Other historians of geology have demonstrated the important role played by earlier forms of antiquarianism in the historicization or temporalization of earth science.[46] Cuvier's rhetoric gives a hint, however, as to what is new in his novel science of geology: the fossil as a "new species" of monument. As practiced by Cuvier, geohistory not only accelerates the develop-

44 Ibid. 1. My emphasis.

45 Noah Heringman, *Sciences of Antiquity: Romantic Antiquarianism, Natural History, and Knowledge Work* (Oxford: Oxford UP 2013)2.

46 The thesis is not new, cf. Cecil Schneer, "The Rise of Historical Geology in the Seventeenth Century," *Isis* 45, no. 3 (September 1954): 256–268, but Heringman has given it its fullest treatment to date. Mainly concerned with the impact of "deep time" on historical thought, Paolo Rossi, *The Dark Abyss of Time: The History of the Earth & the History of Nations from Hooke to Vico*, trans. Lydia G. Cochrane (Chicago: U Chicago P, 1993), also provides a theoretical basis for discussing the intertwining of history and geology in the eighteenth century.

ment of existing historicist paradigms, but also provides, with the fossil, a new "epistemic thing."[47]

Under Cuvier's study, it is the fossil that inaugurates a new approach to history that eventually leads to the art conservation movements of the mid-nineteenth century. In his work on fossil bones, Cuvier provides a model of how "dispersed" fragments could be brought together and reconstructed into a coherent whole. His science of the coexistence of forms, or comparative anatomy, will allow Viollet-le-Duc, for example, to infer that architectural forms, like organs, have functions and undergo an evolution like living beings. While Viollet-le-Duc's analogy is fanciful, it indicates the general principle of Cuvier's impact. With Cuvier, the nineteenth century begins its search for new sciences and techniques (*arts*) to reconstruct past worlds that have been ruined and fragmented by time. For as the paleontologist writes, the reconstruction of fossil bones is also necessary for the broader project of a "history of the planet," the task of geology. In the following pages of his *Discours*, he turns to the example of sea fossils found on the tops of mountains as evidence of the catastrophes and revolutions of earth history. Even "l'observation la plus superficielle" [the most superficial observation] suggests that something as monumental as the ocean has undergone "au moins un changement, soit en étendue, soit en situation" [at least one change, either in its extent, or in position].[48]

In Cuvier's historical geology, the fossil is the key to the past.[49] He can observe extinct life forms, reconstruct skeletal remains, and even entire living environments. In turn, as a geologist, he can date strata according to their respective fossil compositions, and thus define an epoch as an arrangement of its living beings. The geological stratum, as a repository for fossil remains, is the earth's own time capsule. At the same time, such strata are not immobile. Through erosion and elevation, through the push and pull of the earth's movements, these time capsules become dispersed. In putting together the broken and scattered fragments of the past, the geologist must also attempt to distinguish between what is broken and what scattered.

---

47 On the concept of "epistemic thing," see Hans-Jörg Rheinberger, *Toward a History of Epistemic Things: Synthesizing Proteins in the Test Tube* (Stanford, Calif.: Stanford UP, 1997). For a similar, yet culturally broader conception of "objects of knowledge," see the example of "Kepler's moon" in Vogl, "Poetologie des Wissens," 49. Foucault discusses fossils and monsters, both as disruptive forces for natural-historical taxonomies, cf. Foucault, *The Order of Things*, 50–56.

48 Cuvier, *Recherches sur les ossements fossiles*, 6.

49 For a detailed discussion of this new method, see Martin Rudwick, *Georges Cuvier, Fossil Bones, and Geological Catastrophes* (Chicago: U Chicago P, 1997).

By 1830, at least in Charles Lyell's English usage, the "historian" has replaced the "antiquarian" at the top of the hierarchy of professions that deal with the past, but the relationship between (historical) geology and (human) history is as strong as ever. In Lyell's definition of geology on the first page of his *Principles*, the keyword that identifies the legacy of James Hutton's and Georges Cuvier's historicization of the field is *successive:* "Geology is the science which investigates the successive changes that have taken place in the organic and inorganic kingdoms of nature."[50] The notion of a "succession of events" and the inquiry into the "causes of these changes" unites geology with history, and separates both from the search for origins that defined the eighteenth century's approach to the past. The analogy between history and geology is not just illustrative; it is programmatic. The "changes" and "causes" in earth history are *plural*, and like the historian, the geologist should be interested in what *has in fact* "taken place."[51] The "temporalization of nature" implicit in earlier geological theories thus gives way to a *historicization of nature.*[52]

In the prominent analogy between geology and history, moreover, Lyell also brings geology into the age of historicism, where history and historical knowledge become essential to knowledge as such. It is significant that Lyell, famous for the maxim "the present is the key to the past," nonetheless begins the *Principles* with the typical historicist argument that the *past* is the key to the present: "As the present condition of nations is the result of many antecedent changes, some extremely remote and others recent, some gradual, others sudden and violent, so the state of the natural world is the result of a long succession of events, and if we would enlarge our experience of the present economy of nature, we must investigate the effects of her operations in former epochs."[53] It is not hard to imagine Charles Darwin, who owed a great debt to Lyell's work in geology and whose theory of evolution gave rise to ecology, a new science of the "present economy of nature," finding particular inspiration in these words. For his part, Lyell hoped to induce zoologists and biologists "to sacrifice time in exploring the imperfect remains of lost species of animals and plants [out of] a desire of promoting *geology* by such investigations."[54]

As a *scientia universalis* of the nineteenth century, history is connected "with all branches of knowledge, whereby any insight into human affairs, or into the

---

50 Lyell, *Principles of Geology*, 1.

51 Ibid.

52 So the title phrase from Peter Matussek, ed., *Goethe und die Verzeitlichung der Natur* (Munich: Beck, 1998).

53 Lyell, *Principles of Geology*, 1.

54 Ibid. 4.

moral and intellectual nature of man, can be obtained."[55] Likewise, geology is a central node in a network of natural sciences which includes "chemistry, natural philosophy, mineralogy, zoology, comparative anatomy, botany; in short, in every science relating to organic and inorganic nature."[56] The geologist thus seeks the aid of scientists from all these fields, and even appeals to the larger group of antiquarians, that they might turn their passion for collecting to the "relic[s]"[57] of the earth: "According to different opportunities, tastes, and talents of individuals, they may employ themselves in collecting particular kinds of minerals, rocks, or organic remains, and these, when well examined and explained, afford data to the geologist, as do coins, medals and inscriptions to the historian."[58] Here too, Lyell's call did not go unheeded. Nineteenth century geologists could rely on many avid readers to agree that, compared to the revelations of political and social history, the "connexions brought to light, when we carry back our researches into the history of nature [are] far more astonishing and unexpected."[59] As Cuvier established two decades prior, and Stifter will reconfirm two decades later, the geological age is a time of superlatives: "L'histoire ancienne du globe [...] est d'ailleurs par elle-même l'un des objets les plus curieux qui puissent fixer l'attention des hommes éclairés" [the ancient history of the earth is, by the way, itself one of the most curious objects that can fix the attention of enlightened minds].[60]

On the Mountaintop: Gazing into the Depths of Time

Towards the end of the eighteenth century, the aesthetic category most associated with the fascination, curiosity, and astonishment at geological deep time was the sublime. Indeed, in Immanuel Kant's theory of the sublime, mountains and mountain ranges qualify by virtue of their size; they are "simply big," as Imma-

55 Ibid. 2.

56 Ibid. 3.

57 In addition to his explicit analogies, Lyell frequently employs terms like "relic," "monument," and "ruin" as metaphors that connect earth and human history.

58 Ibid. 4. The theme of collecting is central to Stifter's work as well, cf. Kathrin Maurer, "Adalbert Stifter's Poetics of Collecting: Representing the Past against the Grand Narrative of Academic Historicism," *Modern Austrian Literature* 40, no. 1 (2007): 1–17, who rightly argues that this poetics contrasts with progress-oriented forms of historicism: geohistoricism is not a grand narrative.

59 Ibid. 3.

60 Cuvier, *Recherches sur les ossements fossiles*, 2.

nuel Kant writes: "schlechthin groß."[61] Before the rise of widespread mountain tourism, moreover, rocky landscapes above the tree line were seen as dangerous, terrible and horrifying wastelands.[62] An early geological essay by Goethe, "Über den Granit" [On Granite, 1784], however, exemplifies how the mountain also came to represent a particular historical variant of the sublime feeling that one might call the *geological sublime*. This feeling thus not only inspires Goethe to a lifelong engagement with geological research, but it also accompanies nearly every first experience of the depths of geological time since then.[63] At the same time, Goethe's feelings about geology change significantly over his lifetime. By the 1820s, the poet-scientist of the court of Weimar will meet the developments, sketched above as the rise of geohistory, primarily with resignation and bewilderment. Goethe's personal transition from an enthusiastic young geognost to a "superannuated Neptunist" thus provides an excellent case study of how the aesthetics of geology were transformed by earth science's historical turn.[64]

In the geological sublime – and unlike the ocean, for example – mountains begin to suggest to the scientist-poet that their greatness has a temporal as well as a spatial dimension. Goethe wrote *On Granite*, an introduction to a larger, unfinished treatise, after reading Buffon's *Époques de la Nature*.[65] Goethe is sensitive to the poetic impact of Buffon's ground-breaking vision of the earth's past, even as he defends its scientific validity against skeptical critics: "Buffon habe einen Roman geschrieben, welches sehr wohl gesagt ist, weil das ehrsame Publicum alles außerordentliche nur durch den Roman erkennt" [It is said that Buffon has written a novel. Well said indeed, as the honorable public can only rec-

---

61 Immanuel Kant, *Kritik der praktischen Vernunft. Kritik der Urtheilskraft. Akademie Ausgabe V* (Berlin: Königlich Preußische Akademie der Wissenschaften, 1913), 248.

62 On the sublimity of stone, cf. Hartmut Böhme, "Das Steinerne. Anmerkungen zur Theorie des Erhabenen aus dem Blick des 'Menschenfremdesten,'" in *Das Erhabene: Zwischen Grenzerfahrung und Größenwahn*, ed. Christine Pries (Weinheim: VCH, Acta Humaniora, 1989), 119–142.

63 Goethe's geological research is well documented, though without current historical methodology, by Wolf von Engelhardt, *Goethe im Gespräch mit der Erde: Landschaft, Gesteine, Mineralien und Erdgeschichte in seinem Leben und Werk* (Weimar: Böhlau, 2003).

64 Goethe sketched out "Eines verjährten Neptunisten Schlußbekänntniß Abschied von der Geologie" in 1819, cf. *Die Schriften zur Naturwissenschaft (Leopoldina). Abt. I Bd. 11. Aufsätze, Fragemente, Studien zur Naturwissenchaft im Allgemeinen*, ed. Dorothea Kuhn and Wolf von Engelhardt (Weimar: H. Böhlau, 1970), 132.

65 On the composition and genre of the text, see Peter Schnyder, "Grund-Fragen: Goethes Text 'Über den Granit' als 'Ur-ei' der Wissensrepräsentation," in *"Ein Unendliches in Bewegung". Künste und Wissenschaften im medialen Wechselspiel bei Goethe*, ed. Barbara Naumann (Bielefeld: Aisthesis, 2012), 245–263.

ognize something extraordinary when [it] is written in a novel].[66] Not to be outdone, an enthusiastic young Goethe hatched a plan to write his own "Roman über das Weltall" [novel about the universe], a project that never came to fruition, but that did result in a number of remarkable fragmentary texts, including the essay *On Granite.*[67] Just as Buffon's work was a milestone in the discovery of deep time, Goethe's short text is an aesthetic achievement that envisions how the depth of time might be integrated into the familiar aesthetic scene of the mountain panorama.

On Granite transforms a scientific discussion of the rock into an emotional drama. The scene is a mountaintop panorama and the emotion the shudder of the sublime.[68] These are familiar topoi that have accompanied the history of alpinism from Albrecht von Haller (*Die Alpen*, 1729) to Reinhold Messner (*The Crystal Horizon*, 1989). Goethe's *On Granite*, however, transforms the typical sublime feeling produced by the mountain landscape – a spatial grandeur – to suggest the *temporal grandeur* of deep time. For granite is not only a rock commonly found on mountain peaks; Goethe's text also apostrophizes granite as the *oldest* type of rock: "ihr ältesten, würdigsten Denkmäler der Zeit" [you, the most ancient and worthiest monuments of time].[69] This understanding of granite most likely comes directly from Buffon, who believed that granite emerged from the "primitive fire" (*feu primitif*) as the earth first began to cool and solidify.[70] Imagined to run without break from the mountaintop to the innermost earth, granite is thus the medium that allows a direct experience of the very beginnings of the world. It is the "fest[e] Boden der Urwelt" [firm foundation of the primal world], the "ältest[e] und ewig[e] Altar" [primal and everlasting altar], where one can bring a sacrifice to "dem Wesen aller Wesen" [the being of all beings].[71]

---

66 Letter to Merck, October 1780, cited in Hans Blumenberg, *Die Lesbarkeit der Welt* (Frankfurt a.M.: Suhrkamp, 1983), 225.

67 On Goethe's "novel about the universe" as a variant of the metaphor 'book of nature,' see Blumenberg, *Lesbarkeit der Welt*, 214–232; by contrast, Annette Graczyk situates the project within the context of the eighteenth-century 'tableau,' *Das literarische Tableau zwischen Kunst und Wissenschaft* (Munich: Fink, 2004), 190–196.

68 Cf. Elizabeth Powers, "The Sublime, 'Über den Granit,' and the Prehistory of Goethe's Science," *Goethe Yearbook* 15 (2008): 35–56.

69 Johann Wolfgang von Goethe, "Granit II," in *Die Schriften zur Naturwissenschaft (Leopoldina). Abt. I Bd. 11*, 10–14; "On Granite (1784)," in *The Essential Goethe*, ed. Matthew Bell (Princeton, NJ: Princeton UP, 2016), 914.

70 Cf. Comte de Buffon (Georges-Louis Leclerc), "Du granit," in *Histoire naturelle des minéraux*, vol. 1 (Paris: Imprimerie Royale, 1783), 96–129.

71 Goethe, "Granit II," 12 [914].

Through the medium of granite, Goethe is able to move from the sublime mountain panorama, "Auf einem hohen nackten Gipfel sitzend und eine weite Gegend überschauend" [as I stand high atop a barren peak and survey the wide expanse below], to an overview that is not just spatial, but temporal.[72] From a description of the mountain panorama, the text gradually transitions to a narrative of earth history, which, of course, includes "scenes of devastation" that threaten to destabilize the subject of the narrative: "In der Ferne heben sich tobende Vulkane in die Höhe, sie scheinen der Welt den Untergang zu drohen" [Raging volcanoes rise up in the distance, seeming to threaten the world with destruction].[73] This is a vivid expression of the geological 'outrage,' of the thought that our entire world could be destroyed just as it was created, in a fire. Yet the discourse of the sublime not only makes it possible to imagine time and power on a grand and terrifying scale; it also provides techniques of psychic relief. One of these is as banal as distance: the speaking subject remains safe and "unerschüttert [...] indes die Bewohner der fernen Ufer und Inseln unter dem untreuen Boden begraben werden" [unshaken while those who live on distant shores and islands are buried beneath the faithless land].[74] In other words, the shocks of a catastrophe like the earthquake in Lisbon in 1755 may have reverberated throughout Europe, but only the locals were buried.

Yet as we have seen, the observation of *granite* cliffs, in particular, produces a sublime feeling because it brings the subject closer to the *truth* of the earth's past, the "ältesten ersten tiefsten Gefühlen der Wahrheit" [deepest, oldest, most elemental feeling for the truth].[75] In this respect, Goethe's mountaintop scene is comparable to Kant's famous analogy in the *Critique of Practical Reason* between "der bestirnte Himmel über mir" [the starry heavens above] and "das moralische Gesetz in mir" [the moral law within].[76] The "granite below" represents the depth of time just as the starry sky represents the expanse of space. The only problem with this comparison is that there is no law to explain earth history as elegantly as Newton's law of gravity explains the movements of planets and stars.[77] Goethe's text names the obstacle quite clearly:

72 Ibid.

73 Ibid. 13 [915].

74 Ibid.

75 Ibid. 12 [914].

76 Immanuel Kant, *Kritik der praktischen Vernunft, Akademie Ausgabe V*, 161.

77 Cuvier still expresses the hope that there will one day be a geological Newton – perhaps even himself – in 1812, *Recherches sur les ossements fossiles*, 3.

> Hier ruhst du unmittelbar auf einem Grunde, der bis zu den tiefsten Orten der Erde hinreicht, keine neuere Schicht, keine aufgehäufte zusammengeschwemmte Trümmer haben sich zwischen dich und den festen Boden der Urwelt gelegt, du gehst nicht wie in jenen fruchtbaren schönen Tälern über ein anhaltendes Grab, diese Gipfel haben nichts Lebendiges erzeugt und nichts Lebendiges verschlungen, sie sind vor allem Leben und über alles Leben.
>
> Here you stand upon ground which reaches right down into the deepest recesses of the Earth; no younger strata, no pile of alluvial debris comes between you and the firm foundation of the primal world. What you tread here is not the perpetual grave of those beautiful, fruitful valleys; these peaks have never given birth to a living being and have never devoured a living being, for they are before all life and above all life.[78]

Except on the barren summit, what lies between the self-dialogizing speaker of these lines and the primordial world is "life," whose very fertile productivity is transformed under the geological gaze into a "perpetual grave." Life is not only more recent than the original stone, "younger strata," but it is also constantly producing new layers. The phrases "pile of" and "alluvial" intensify the key word here: debris, or *Trümmer.* Rubble means "Trümmer, Unordnung und Zerstörung" [ruin, chaos, and destruction], not the "firm foundation" that the speaker is seeking.[79] The order of earth history is guaranteed here by the primacy of origins, for if one has a theory that describes the formation of the oldest rock – be it the primitive ocean of Neptunism or Buffon's "primitive fire" – then all the rest is secondary, because the earth has its ultimate logic.

Without rigorous historicization, moreover, the theory of the earth does not necessarily need to come to terms with some of the more radical implications of geological deep time. Circa 1800, among early German Romantics, it remains common to imagine deep time as an *Urzeit* that played by its own rules and has little effect on our present. Buffon's theory of the cooling earth suggests this model. A student of Abraham Gottlob Werner, Friedrich von Hardenberg (Novalis) also subscribes to the idea that the earth was once much more chaotic than it is today: "Wie ruhig und friedfertig, wie mild und klar ist gegen diese gewaltsamen, riesenmäßigen Zeiten, die heutige Natur! Und das furchtbarste Gewitter, das entsetzlichste Erdbeben in unseren Tagen ist nur ein schwacher Nachhall jener grausenvollen Geburtswehen" [How quiet and peaceful, how mild and dear is our present nature, when compared with violent and gigantic times! The mightiest tempests, the most terrible earthquakes of our day, are but weak ech-

78 Goethe, "Granit II," 12 [914].
79 Ibid. 13 [915].

oes of the throes of that first birth].[80] The directional theory of earth history also aligns with teleological narratives of *Bildung* or universal progress. In his famous *Bildungsroman, Heinrich von Ofterdingen*, Novalis thus symbolizes the *Urwelt* as a cave. If Novalis calls the chaotic tumult of the primitive earth its birthing pains, the cave in the protagonist Heinrich's dream, which initiates the young man's quest for the famous blue flower, takes on the characteristics of a uterus. Geology thus inspires the young poet to dream of his birth, or rather, to return to his birthplace and be reborn with a new knowledge of the past. Furthermore, the earth's directional development resembles the development of the poet: "[Die Natur] nähert sich dem Menschen, und wenn sie ehemals ein wildgebärender Fels war, so ist sie jetzt eine stille, treibende Pflanze, eine stumme menschliche Künstlerin" [Nature approaches man; and if she were once an uncouthly teeming rock, then now she is a quietly thriving plant, a silent human artist].[81]

As Peter Schnyder has demonstrated, Goethe also clung stubbornly to the idea of the earth's present stability.[82] Indeed, in Goethe's interpretation of the famous Serapis Temple in Pozzuoli near Naples, this belief took the form of an unquestioned 'constant' in the scientific sense, as well. When Goethe visited Pozzuoli during his Italian travels in 1787, the Serapis Temple had already acquired a reputation as a geo-archeological puzzle, which only grew in the decades that intervened before Goethe published his findings in 1823. The Temple's columns display well-defined rings or ribbons of marine fossils, which suggest a number of significant geological changes during the historical period since the Greeks constructed it. Goethe's solution to this famous problem involved three steps: first, a volcanic eruption covered the bottom of the columns in ash; second, a pond gathered on this ground, leaving marine fossils; third, both pond and ash receded in later erosion. Less important than this explanation, though, is the fact that Goethe's sketches of these steps are intentionally superimposable. The *bedrock* remains a constant, even as the superficial ground above can change.

---

80 Novalis, *Werke*, ed. Gerhard Schulz (Munich: Beck, 1969), 201; *Henry of Ofterdingen. A Romance*, ed. John Owen (Cambridge, MA: Cambridge Press, 1842), 112.

81 Novalis, *Werke*, 202 [113]. Although these reflections represent the most current geology of the time around 1800, such teleological interpretations of the earth's formation are compatible with mythological cosmogony. Indeed, Novalis's assimilation of science, mythology and poetry is programmatic. New discoveries in geology do not contradict, but rather confirm ancient myth: "wenigstens dürfte man die alten Sagen von einem Riesenvolke dann keiner Erdichtungen zeihen," 201.

82 Peter Schnyder, "Die Dynamisierung des Statischen. Geologisches Wissen bei Goethe und Stifter," *Zeitschrift für Germanistik* 19, no. 3 (2009): 540–555.

Charles Lyell, who chose the Serapis Temple to adorn the frontispiece of his *Principles of Geology*, not only provides a definitive explanation of the phenomenon at work, but also points to the "epistemological obstacle" (Bachelard) that had prevented such insights in the past: "It is time that the geologist should in some degree overcome those first and natural impressions which induced the poets of old to select the rock as the emblem of firmness – the sea as the image of inconstancy."[83] Again, less important than the 'Vulcanist' theory of uplift is the "reevaluation of values" (Schnyder) that it implies. Indeed, Lyell himself reacts to the radicalism of his actualist history with a compensatory theory of geological equilibrium. For every rise in the earth's crust there must be a corresponding collapse, for each accumulated sediment a corresponding erosion. By contrast, Goethe, if he was forced to accept certain evidence of Vulcanism, could only respond to Alexander von Humboldt's *Fragmens de géologie et de climatologie Asiatiques* (1831) with characteristic ironic resignation; to imagine the Himalayas rising from the ground, his "Zerebalsystem müßte ganz umorganisiert werden" [cerebral system would have to be completely reorganized].[84]

Lyell's reference to "the poets of old" suggests that these scientific developments were accompanied by changes in the aesthetics of geology. Each of the two main paradigms of geological aesthetics in question here is associated with privileged object or motif. For the eighteenth-century theories of the earth, this motif is the mountain, or more specifically, its *granite* core.[85] For nineteenth century historical geology, it is the fossil. Although I argue that the fossil aesthetic emerges later, the mountain aesthetic does not merely disappear.[86] It remains a central theme in Stifter's *Indian Summer*, and today we still call the view of an imposing mountain range sublime. Yet the sublimity of the inorganic tends to overshadow and obscure the importance of fossils, as my reading of Goethe's *On Granite* has shown. Aside from dinosaur skeletons that only impose when reconstructed, the organic remains of the deep past tend to occupy, in the popular imagination, the place of the souvenir. And if the nautilus is the most

---

83 Lyell, *Principles of Geology*, 156. On the concept of 'epistemological obstacle,' see Gaston Bachelard, *The Formation of the Scientific Mind: A Contribution to a Psychoanalysis of Objective Knowledge*, trans. Mary McAllester Jones (Manchester: Clinamen, 2002).

84 Goethe to Zelter on October 5, 1831, cited in Schnyder, "Dynamisierung des Statischen," 550.

85 Cf. Rüdiger Görner, "Granit. Zur Poesie eines Gesteins," *Aurora* 53 (1993): 126 – 138.

86 Similarly, it is worth remembering that geologists continue to research the earth's earliest days – i.e., the accretion of solar dust under the forces of gravity, the formation of the moon and other heavy meteoric bombardment in the aptly named 'Hadean' eon. By definition, however, this eon lacks a fossil record, which means that such research has little in common with the mainstream of historical geology.

cherished of such souvenirs, it is not only because of its beauty, but also because of its wholeness.

The provocation of earth history, however, is that it does *not* have any logic so simple, or as Goethe will later write in his 1829 novel *Wilhelm Meisters Wanderjahre* [*Wilhelm Meister's Journeyman Years*]: "Daher kann keine Geologie geben, denn die Vernunft hat hier nichts zu tun" [Hence there can be no geology, for reason has no role here].[87] Hans Blumenberg argues that this encounter with the earth as the limit of reason leads Goethe to a reappraisal of his project for a "novel about the universe." On the one hand, it is no longer possible to tell the story of the earth because stone is silent: "Steine sind stumme Lehrer, sie machen den Beobachter stumm, und das Beste was man von ihnen lernt ist nicht mitzuteilen" [Rocks are mute teachers; they render the observer mute, and the best thing one can learn from them is to keep one's counsel].[88] Following the lesson of this aphorism, Montan, the geologist or "observer" of rocks in Goethe's novel, exemplifies the novel's poetic strategy of *Entsagung* or resignation. Goethe, however, says a great deal about geology in this novel. At the mountain festival, or *Bergfest*, miners exchange a range of opinions about the origin of the earth that covers almost all the major theories of the day, from Vulcanism and Neptunism to meteorites and nascent glaciation theories. The expert, Montan, remains silent, however, refusing to pronounce judgement on these theories because none of them expresses the "the being of all beings" that seemed so near when Goethe wrote his essay *On Granite.*

On the other hand, the silence of stone is also the precondition of its poetic character. In Blumenberg's reading, the failure of reason is also "the discovery of the Romantic region, the depth of the earth, the Cave."[89] The metaphors of crystallization (*Kristallisationslust*) and coagulation (*Gerinnen*, primarily of marble) can be seen as the beginnings of a new poetry: "Wenn man von Uranfängen spricht, so sollte man uranfänglich reden, das heißt dichterisch" [When one speaks about primal beginnings one should speak primordially, that is, poetically].[90] This story of German "Romantic Rocks" leads from Goethe to Novalis,

87 Johann Wolfgang von Goethe, *Wilhelm Meisters Wanderjahre. Sämtliche Werke*, ed. Gerhard Neumann and Hans-Georg Dewitz, vol. 10 (Frankfurt a. M.: Deutscher Klassiker Verlag, 1989), 582; *The Collected Works*, ed. Jane K. Brown, trans. Jan van Heurck and Krishna Winston, vol. 10 (Princeton, NJ: Princeton UP, 1995), 311. Also cited in Blumenberg, *Lesbarkeit der Welt*, 227.

88 Ibid. 762 [429].

89 Blumenberg, *Lesbarkeit der Welt*, 227.

90 Goethe as cited in Blumenberg, ibid.

Tieck, and indeed to Stifter, who remains fascinated with the crystalline forms shared by rock and ice, and the milky coagulation of marble.[91]

Although inorganic rocks are silent, which makes their 'message' poetic, in the sense of non-discursive, there are objects in the broader field of geology that *do* speak. Turning his attention to the organic remains that he so willfully ignored in 1784, the older Goethe writes: "Wenn ich ein zerstreutes Gerippe finde, so kann ich es zusammenlesen und aufstellen; denn hier spricht die ewige Vernunft durch ein Analogon zu mir, und wenn es das Riesenfaultier wäre" [When I find the fragments of a skeleton, I can gather them and put them together. For here eternal reason speaks to me through an analogue, even if the skeleton be a giant sloth].[92] For Blumenberg, the fact that the organic fossil speaks while the inorganic rock is silent explains how there can literally be no geo-logy, no logical discourse of the earth.[93] A scientific-historical interpretation sees something different in the reference to "fragments of a skeleton" and the reconstruction of a "giant sloth." In the 1820s, when Goethe wrote these lines, such phrases pointed clearly to Cuvier's comparative anatomy of fossil bones and new logic of geology as geohistory.

In a certain sense, this logic was always there, an unknown only in the sense that it was repressed. In the essay *On Granite*, "ruin, chaos, and destruction" refer not only to the history of the earth, but also to the history of geology, which the young Goethe finds in the "library" or "Studierstube." Or rather, Goethe finds both the disorder of earth history – "Nothing here is in its primal, ancient state" – and the disorder of scholarly opinion.[94] The commonplace criticism of book culture (i.e. *Gelehrsamkeit*) conveniently covers up the underlying disorder of the 'book of nature.' In the *Journeyman Years* novel, Montan indulges a similar critique of books and writing:

> Buchstaben mögen eine schöne Sache sein, und doch sind sie unzulänglich, die Töne auszudrücken; Töne können wir nicht entbehren und doch sind sie bei weitem nicht hinreichend, den eigentlichen Sinn verlauten zu lassen; am Ende kleben wir am Buchstaben und am Ton, und sind nicht besser dran, als wenn wir sie ganz entbehrten; was wir mitteilen, was uns überliefert wird, ist immer nur das Gemeinste, der Mühe gar nicht wert.

---

91 To adapt the title of Noah Heringman's study of British Romanticism, *Romantic Rocks, Aesthetic Geology* (Ithaca, NY: Cornell UP, 2004).

92 Goethe, *Wilhelm Meisters Wanderjahre*, 582 [311].

93 "Das Organische redend, das Anorganische stumm, damit erschließt sich die Wörtlichkeit, mit der genommen sein will, daß es keine Geo'logie' geben kann," Blumenberg, *Lesbarkeit der Welt*, 228–229.

94 Goethe, "On Granite (1784)," 915; "fast nichts in seiner alten Lage," "Granit II," 13.

> The letters of the alphabet are fine things, but they are inadequate to express sounds; we cannot do without sounds, and yet they are by no means sufficient to express actual meaning; in the end, we cling to letters and sounds, and we are no better off than if we did without them altogether. Everything we communicate, everything that is transmitted to us, is always only the obvious, not worth the effort at all.[95]

This invective against books is Montan's straw man, a preface to the justification of his research in geology. Yet in explaining this interest, he uses the very same terms, the same metaphor of writing, now valued positively. Montan treats the rock features, "these very fissures and crevasses as letters."[96] Indeed, Montan's terms are so ambivalent that one can easily imagine extending the metaphor: the "fissures" (*Spalten*) and "crevasses" (*Risse*), if one translates geologically, become "columns" and "tears," to translate philologically. As such, the letter already appears to have two sides. It is the semiotic difference that allows the "text" to make sense, and thus it is a legible gap, or a gap that produces legibility; but, as *tear*, it is an absolute gap, an irreparable hole that only produces fragments.

Understandably, Wilhelm responds to Montan's analogy with some skepticism: "es scheint mir ein weitläufiges Alphabet zu sein" [it seems to me a rather diffuse alphabet].[97] The larger the alphabet, the more illegible the script. In response to this suggestion that the earth's writing may be as difficult to decipher, if not more, than human writing, Montan doubles down on his analogy. He insists that at the very least, the writing of nature is guaranteed to be authentic:

> Die Natur hat nur eine Schrift, und ich brauche mich nicht mit so vielen Kritzeleien herumschleppen. Hier darf ich nicht fürchten, wie wohl geschieht, wenn ich mich lange und liebevoll mit einem Pergament abgegeben habe, daß ein scharfer Kritikus kommt und mir versichert, das alles sei nur untergeschoben.
>
> Nature has only one script, and I do not need to load myself down with all sorts of scribblings. Here I have no fear, as can happen if I work long and lovingly on some parchment, that some clever critic will come along and assure me that I have [been reading an inauthentic text – TA].[98]

95 Goethe, *Wilhelm Meisters Wanderjahre*, 292 [116].
96 Ibid.
97 Ibid.
98 Krishna Wilson translates, "that I have merely read my own meaning into it," which suggests, incorrectly, that the problem here is the reader's subjectivity, not the authenticity of the source, ibid.

For Montan, nature still only has one text or code. As Blumenberg writes, this attitude reflects Montan's belief in "the direct experience of solidness" that refuses to seek "a risk [...] in this configuration that reaches deeply into human affairs."[99] The risk is evident in the double meaning of the word "untergeschoben." In the context of written evidence, *untergeschoben* means "fake" or "inauthentic," but for the geologist, what is *untergeschoben* is merely "pushed under" or "inserted," an immanently natural process.

By no means does this make the earth's pushing and shoving less problematic for the geologist, however. In fact, the varied mechanical processes of geological change are precisely what makes grandiose theorizing as difficult in 1829 (a decade after Goethe's "Abschied von der Geologie" [Departure from Geology] in 1819) as it was in 1784, when Goethe first encounters the rubble of geological theory.[100] In the history of the world's legibility, the advent of geohistory thus represents a new epoch for two reasons: first, the legibility of the natural world is confirmed in a significant new way. Stones, rock formations, and fossils become the material archive of their own history, comparable to letters and texts. But, second, the story they tell does not have the unity or coherency of a book. Nature is no longer a *system* that can be depicted in an encyclopedia, but rather a dynamic, open-ended history whose paths are contingent and varied. Crystallization, sedimentation, metamorphosis: such rock-producing processes leave behind evidence that can be traced back to the process itself. But erosion, eruption, and subterranean (or tectonic) forces, complicate the story. Whether Montan knows it or not, what geology in the nineteenth century requires, and will soon have, is "sharp critics," who carry around and compare the many scribbles that the earth bears on its ragged surface.

Ex uno plura: The Archive of Earth History

The horizon of this generalized earth history (itself contained in turn by the history of the solar system, the galaxy, and the universe) finds its most condensed metaphorical expression in Stifter's word "interpolation" or *Einschiebsel*, an etymologically related counterpoint to Goethe's *untergeschoben*. When nature only has "one script," it is easier to distinguish the authentic from the fake, unlike

99 "der unmittelbaren Erfahrung von Festigkeit [...] ein Risiko wird spürbar in dieser Konfiguration, das tief ins Menschliche hineinreicht," Blumenberg, *Lesbarkeit der Welt*, 225.
100 See note 64.

when the earth's archive contains, as it does in Stifter's novel, "millions of documents:"

> Die Quellen zur Geschichte der Erde bewahrt sie selber wie in einem Schriftengewölbe in ihrem Innern auf, Quellen, die vielleicht in Millionen Urkunden niedergelegt sind, und bei denen es nur darauf ankömmt, daß wir sie lesen lernen, und sie durch Eifer und Rechthaberei nicht verfälschen.
>
> The Earth itself preserves the sources of this history in its innermost parts just as in a room for records, sources inscribed in perhaps millions of documents; it is only a matter of our learning to read and not falsify them by eagerness or obstinacy.[101]

The metaphor of the archival vault, the *Schriftengewölbe*, continues the tradition of the 'book of nature,' yet robs it of one of its most salient qualities, the implied unity of the single book. Instead, the geologist stands before an immense archive with millions of sources.[102] If the passage does not question the integrity of the archive, it does suggest that there may be more at stake in "falsifying" the records than appears at first glance. At least, if we read *Einschiebsel* with its more pejorative connotations, then the "interpolation" of human history has already falsified the history of the earth.

In addition to the metaphor of the *Einschiebsel*, paradigm of the nineteenth century as a *geological age*, Stifter's novel also provides a figure of thought that exemplifies the spirit of this age's *restorative poetics*. This thought, expressed by the geologist's older friend and mentor the Freiherr von Risach, suggests that the *Nachsommer*, the "post-summer," can or even must do without the summer itself: "So leben wir in Glück und Stettigkeit gleichsam einen Nachsommer ohne vorhergegangenen Sommer" [Thus, we are living in happiness and with a sense of constancy as if in an Indian Summer without the preceding summer].[103] What is an "Indian summer without the preceding summer?" The structure of Risach's statement suggests an 'after' without a 'before,' that is, a state of pure belatedness or posteriority – or, at the very least, the 'before' somehow did not take place; it is a curious non-place or *non-lieu*. The gambit of this book is that this non-lieu is identical to the past as it is reconstructed by restorative practices. Stifter's curious formulation is all the more striking, I argue, when paired with

101 HKG 4.2, 32 [192].

102 On astronomical space and geological time as media of communication, see John Durham Peters, "Space, Time, and Communications Theory," *Canadian Journal of Communication* 28 (2003): 397–411.

103 HKG 4.3, 224 [445]. Here (*Stettigkeit*) as elsewhere, Stifter's orthography is so frequently unusual that I have not marked unconventional spellings with *sic*.

Viollet-le-Duc's famous description of the goal of architectural restoration as "un état complet qui peut n'avoir jamais existé à un moment donné" [a condition of completeness that may never have existed at any given time].[104] The methods of geological reconstruction, which inspire Viollet-le-Duc's practice of restoration, provide entirely new and invigorating possibilities for reestablishing this "condition of completeness." For critics of this fanciful vision of a restoration more complete than the 'original' and of historicist culture more generally, the past suffers from a fundamental lack with respect to the present. We can never really know history 'as it really was' because we can never relive it. In the mind of the geohistoricist, however, the fragmented or degraded way in which the past presents itself in its material remains becomes a vector of desire. In this way, geology and its knowledge of erosion and destruction also *motivates* the rise of restorative historicism. If the nineteenth century begins with a fascination with the romance of ruins – which Diderot ironically sums up with the imperative, "Il faut ruiner un palais pour en faire un objet d'intérêt" [to make a palace interesting, one must reduce it to ruins][105] – its historicist core is driven by the dream of recreating, scientifically and imaginatively, the historical worlds which those ruins and fragments of the past represent. In other words, to make a palace interesting, one must reconstruct it.

Because of the all-too-obvious *lack* inherent in recreated pasts – be it in a restored chateau, a panorama, a historical novel, or a museum reconstruction of a dinosaur's skeleton or a computer-enhanced reconstruction of its flesh – the belief in their power to invoke the past has often been considered naïve. Nietzsche struck at the heart of his period's "antiquarian spirit" when he mocked the antiquarian's excessive identification with the past.[106] Yet the geologist, almost by definition, is immune to the antiquarian's identificatory fantasies, for the worlds she discovers are without humans. In Stifter's literary work, there is thus a constant tension between the forces of erosion and the efforts of reconstruction.[107] In the *Nachsommer*, Stifter integrates this tension into the arc of a

---

104 Viollet-le-Duc, "Restauration," 14 [9]. Translation modified from "could never have existed" since Viollet-le-Duc's phrase only suggests that the historical existence of a "complete condition" is uncertain.

105 Denis Diderot, "Salon de 1767," in *Oeuvres complètes, vol. 16*, ed. Else Marie Bukdahl, Michel Delon, and Annette Lorenceau (Paris: Hermann, 1990), 348.

106 Cf. Friedrich Nietzsche, "On the Utility and Liability of History for Life," in *Unfashionable Observations*, trans. Richard T. Gray (Stanford: Stanford UP, 2003), 83–169, which I discuss in Chapter 4.

107 Stifter also imagines culture as an erosive force, cf. Sabine Schneider, "Kulturerosionen. Stifters prekäre geologische Übertragungen," in *Figuren der Übertragung*, ed. Gamper and Wagner, 249–269.

Bildungsroman: the young geologist learns to reconstruct scientifically, but he also learns to assume his place within a cultural tradition, which itself can only reproduce itself via repetition and reconstruction.

Neither geology nor restoration is a traditional way of dealing with the past. As Viollet-le-Duc claims, "Le mot et la chose sont modernes" [The term Restoration and the thing itself are both modern].[108] In one sense, this modernity is merely empirical; just as geology came into its own (as historical geology) in the early nineteenth century, so too did modern conservation-restoration emerge during Viollet-le-Duc and Stifter's lifetimes. At the same time, historicizing discourses are modern in a broader sense. Like all modern discourse, historicism insists on a fundamental break with the past, drawing a distinction between the ancient and the modern. But historicism goes a step further in suggesting that such fundamental breaks are an ineluctable law of history. In other words, one need not try to be modern to no longer be part of the past, for, despite one's best efforts to revive the past, the past proves irretrievably distant. The insight into the world's historicity is as modern as the imperative to innovate – the difference perhaps being that the one looks back while the other forward.

In the writings of Stifter and Viollet-le-Duc, "restoration" is a compromise, a hybrid form that acknowledges its modern provenance while seeking historical authenticity. The aesthetic practices of geohistoricism certainly seek to stabilize historical change. Restoration seeks to repair and preserve the material past, and thus emphasizes its durability. It tends to fetishize the most durable – the massive stone of a cathedral, the marble of a statue, the jewel of a ring. Nevertheless, geological knowledge undermines the symbolic meaning of stone as the emblem of durability as much as it calls into question the value of the term "ancient" as applied to human history. If stone is transient and the ancient in fact recent, then our world has always already been modern. In other words, a restorative poetics values *kairos* as much as it does stability: the marble statue is most beautiful when illuminated by a lightning storm, and the carefully reconstructed house bears a crown of ephemeral roses.

This book thus reads Stifter's *Indian Summer*, the source of these examples, as the poetological core of his work as a literary writer, painter, conservationist, and (amateur) scientist; that is, as his most elaborate and mature reflection on a set of interrelated cultural approaches to history which characterized not only his own search for historical meaning, but also that of his contemporaries. Key to this reading are the rich similarities between Stifter's theory of *Wiederherstellung* and Viollet-le-Duc's *restauration*, though there is an important difference

108 Viollet-le-Duc, "Restauration," 14 [9].

between them: Stifter's 'restorations' take place first and foremost in the medium of literature. This difference helps explain the tension between words and things that has been the focus of research on Stifter – and German Realism in general – in recent decades.[109] With the advent of geology and archeology, namely, material things acquire a new historical value: they *speak* in addition to, or in the place of, words. The assumption that words can invoke things, and that things in turn can function as transparent signs, is central to the poetics we call 'realist.' Roland Barthes famously calls the transparency of words "referential illusion" and the transparency of things "reality effect."[110] But if Viollet-le-Duc is correct and philology also participates in the spirit of geohistoricism (via the Lachmannian reconstruction of fragmented manuscripts or the Schlegelian "comparative anatomy" of languages), then the materiality of literature itself – the historicity of words and writing techniques – becomes a pressing concern.[111] Stifter's historical novel, *Witiko*, for example, not only invokes the material world of the thirteenth century through historical objects (weapons, clothing, and infamously, an anachronistic fork), but also simulates oral-language codes of the time, all while revising its own historical condition of existence *as literature* by re-dramatizing the origin of the epic genre as exemplified by the *Nibelungenlied* – a self-reflective plot that Cornelia Blasberg has aptly called "erschriebene Tradition" [tradition acquired through writing].[112]

Taking the scientific situation inaugurated by the works of Cuvier, Lyell and other geologists in the 1820s and 1830s as its starting point, Chapter 1 thus begins with Stifter's early literary works, written in the bustling metropolis of Vienna in the early 1840s. Among these works, Stifter's feuilleton-like urban portraiture in the essay collection *Vienna and the Viennese in Images from Their Life*

109 This tension is central to two now classic studies of Stifter: Christian Begemann, *Die Welt der Zeichen: Stifter-Lektüren* (Stuttgart: Metzler, 1995); and Eva Geulen, *Worthörig wider Willen: Darstellungsproblematik und Sprachreflexion in der Prosa Adalbert Stifters* (Munich: Iudicium, 1992). The title of a recent essay collection on German realism (in which Stifter is strongly represented) is programmatic: Sabine Schneider and Barbara Hunfeld, eds., *Die Dinge und die Zeichen: Dimensionen des Realistischen in der Erzählliteratur des 19. Jahrhundert* (Würzburg: Königshausen & Neumann, 2008).

110 Roland Barthes, "The Reality Effect," in *The Rustle of Language* (Berkeley: U of California P, 1989), 148.

111 Cf. Sebastiano Timpanaro, *The Genesis of Lachmann's Method*, trans. Glenn W. Most (Chicago: U Chicago P, 2005); Friedrich Schlegel, "Über die Sprache und Weisheit der Inder. 1808," in *Kritische Friedrich-Schlegel-Ausgabe*, ed. Ernst Behler, vol. 8, (Paderborn: Schöningh, 1975), 137.

112 Cornelia Blasberg, *Erschriebene Tradition: Adalbert Stifter oder das Erzählen im Zeichen verlorener Geschichten* (Freiburg: Rombach, 1998).

(1844) receives particular attention. In these finely crafted, yet light and even humorous works, one can discover the broad and deep underpinnings of Stifter's geological thought. I say broad because certain metaphors such as the 'abyss' of time or the 'erosion' of all things, are common enough to belong to everyday discourse of the journals and almanacs in which Stifter first published his essays and novellas. At the same time, this book argues, Stifter, whose interest in natural science from his student days onward is well documented, continues to dig deeper into geological topics over his career, as he writes the novellas in *Studies* and *Many-Colored Stones* and then finally, *Indian Summer.* The general scheme that will become the explicit topic of that novel, however, is already discernable in Stifter's earliest essays and novellas. It is already apparent that the recognition of inescapable erosion motivates a desire for reconstruction, even as the knowledge required for reconstruction is of the same swath as the insight into processes of erosion.

In the second chapter, then, I turn to an in-depth reading of the novel *Indian Summer:* This 'post-estival' work is Stifter's most fully elaborated literary reflection on the scientific-historical conditions of his epoch – and thus also of itself as a literary work. To this end, the chapter explores the geologist's scientific and aesthetic education and discusses the historical discourses from which they draw. It also analyzes central motifs, such as marble as an aesthetic-geological object, and metaphors, such as the earth's "room of records." Two literary techniques structure the novel as a whole. First, through the character constellation of the student-mentor relationship, Stifter brings together the related disciplines of geology and art history, allowing them to "meet" and exchange ideas. Second, following a technique typical of nineteenth century novels, Stifter embeds the mentor's own life story within the student's first-person narrative. In this way, the past, which is marked by loss and disappointment, is slowly revealed as the "basis" for a fulfilled present. As reconstruction, this present is a repetition and recuperation of the past lost, but it is also irrevocably different from that past: It is attainable only as a late compensation, as *Nachsommer*, or as an uncertain heritage for future generations. In other words, the novel tries to compensate for and work against the entropy of earth history, but it does not pretend to escape it.

The third chapter travels to France to contextualize Stifter's Austrian geohistoricism in a larger European framework. As the introduction has argued, the figure that most resembles Stifter in his understanding of art restoration and its historical genealogy is the architect and restorator Viollet-le-Duc. The chapter thus begins with a discussion of the scientific underpinnings and metaphorical transfers in Viollet-le-Duc's theory of restoration. From start to finish, a fascination with geology both motivated and informed Viollet-le-Duc's vision of architectural restoration. By following this vector through the architect's work, which Viollet-le-Duc himself identifies in the methodological self-reflections that constitute the object of my analysis, I aim to demonstrate, as in the case of Stifter, how the "world of stone" traverses nature and culture and becomes the object of a generalized reconstructive vision and practice.

In the fourth and final chapter, the study returns to Stifter to re-problematize this reconstructive aesthetics with a more critical, media-theoretical perspective. If revivalists imagine a return to the aesthetic heights of the past via immersion in historical *milieu* and imitation of artistic practice, then in its extreme, this practice can only exist as a confined experiment or as an eccentric, countercultural underground – what we observe today in the experimental archeology of Guédelon or in the Society for Creative Anachronism. What is more, as Nietzsche famously argues in his essay "On the Utility and Liability of History for Life," the liability (*Nachteil*) of the "antiquarian spirit" resides in a complete yet false identification with the past and thus a fatal denial of the present and future. By examining the conflicts in the poetic concept behind Stifter's historical novel *Witiko*, this chapter demonstrates how the modern literary medium of the novel cannot sustain the project of restoration as Stifter imagines it. Of course, the novel *can* present, through the conceit of fiction and the techniques of literary realism, an imaginary phantasm of the past – an advantage that places it on par with Viollet-le-Duc's drawings and paintings or today's historical films. What it cannot do – at least not simultaneously – is satisfy the competing demands of the historical sciences (philology, archeology) for historical accuracy and authenticity. The symptoms of this conflict in the novel *Witiko*, I argue, are the archaizing and yet still modernized and thus *simulated* transcriptions of medieval speech, and the inevitable anachronisms in any detailed description of the historical past.

Chapter 1
The Geological Motive: Cultural Critique and Restorative Desire in Stifter's Early Prose

The Metropolitan Prehistory of a "Restorative Project"

In *A New History of German Literature*, Eva Geulen centers her article on Stifter around the date October 11, 1848, the day that, "disturbed by reports from Vienna about the revolution he had initially supported, Adalbert Stifter takes up permanent residence in provincial Linz."[1] This was a significant move in Stifter's personal life, as it marked the end of a twenty-two-year stay in the Habsburg capital. For Geulen, as for most critics, the move also represents a turning point in Stifter's work. Published in 1853, *Bunte Steine* (*Many-Colored Stones*), Stifter's most widely read collection of novellas, appears "in the wake of the failed revolution in 1848" and "programmatically excepts itself from the world of historical upheaval and turns to 'harmless things,' such as the common stones found in the collection's titles: fool's gold, granite, limestone, rock crystal, tourmaline, and bergmilch."[2] *Many-Colored Stones* thus constitutes the center of Stifter's career, which begins with the early narrative *Studies* (from 1840, republished as a book in 1844) that still bear a strong resemblance to the traditions of German Romanticism and particularly Jean Paul, and ends with the more classicizing stylistic rituals of his late novels and novellas.[3] Stifter's life and works thus largely follow the model of the conventional history of German poetic realism, in which the disappointment of 1848 accelerates the turn to a new realist paradigm of "glorification" (*Verklärung*) of the everyday.[4]

---

1 Eva Geulen, "1848, October 11: Tales of a Collector," in *A New History of German Literature*, ed. David E. Wellbery and Judith Ryan (Cambridge, MA: Harvard UP, 2004), 587.

2 Ibid.

3 On the early Stifter's relationship to Romanticism and Jean Paul, see the latest overview by Burkhard Meyer-Sickendiek, "Intertextualität," in *Stifter-Handbuch*, ed. Begemann and Guiriato, 206–209. On Stifter's late style, see Albrecht Koschorke and Andreas Ammer, "Der Text ohne Bedeutung oder die Erstarrung der Angst. Zu Stifters letzter Erzählung 'Der fromme Spruch,'" *DVjS* 61 (1987): 676–719 and Helena Ragg-Kirkby, *Adalbert Stifter's Late Prose: The Mania for Moderation* (Rochester, NY: Camden House, 2000).

4 For a good introduction to this consensus view of German realism and ample historical documentation, see Gerhard Plumpe, *Theorie des bürgerlichen Realismus* (Stuttgart: Reclam, 1985). Paul Fleming offers a compelling knowledge-historical approach in *Exemplarity and Mediocrity: The Art of the Average from Bourgeois Tragedy to Realism* (Stanford: Stanford UP, 2008).

https://doi.org/10.1515/9783110685534-003

If this narrative accurately describes Stifter's place within literary history, it also relies on the commonplace significance of the places in question: Linz is provincial compared to cosmopolitan Vienna. According to this topical logic, the stones likewise belong to the provinces, or more accurately, to nature. In Geulen's account, Stifter is the quintessential *collector*, whose "writing amounts to excavating, collecting, and ordering the world of things in such a way that, ideally, nothing is left out or lost. The only concept of nature that could sustain this project," she adds, "is the early Enlightenment notion of physicotheology, the belief in natural history as the ordered unfolding of a meaningful universe."[5] Yet as recent research on Stifter's knowledge of geology has demonstrated, there is indeed another vision of nature in his work. This vision does not erase the physicotheological tendencies of Stifter's reasoning, but instead resituates him among, or perhaps even at the unexpected and unexplored forefront of those "contemporaries wrestl[ing] with the fact that the older religious faith in nature as God's open book had become incompatible with modern scientific views."[6]

The geological knowledge in Stifter's literary works is more than merely proof that the author is scientifically up-to-date;[7] it demands that we reconsider Stifter's "restorative project."[8] Geulen claims that Stifter embarks on a "restorative project" because the concept of nature that can support his values is irretrievably lost to history. This argument implies not only that history is linear, if not progressive (Geulen mentions Darwin here), but also that "restoration" aims to *undo* this progress.[9] Of course, this is precisely the meaning of "restoration" in the political realm – a connotation that holds, though perhaps less

5 Geulen, "Tales of a Collector," 588.

6 Ibid. On Stifter's concept(s) of nature, see Christian Begemann, "Metaphysik und Empirie: Konkurrierende Naturkonzepte im Werk Adalbert Stifters," in *Wissen in Literatur im 19. Jahrhundert*, ed. Lutz Danneberg and Friedrich Vollhardt (Tübingen: Niemeyer, 2002), 92–127.

7 The extent to which this question has been nonetheless a fruitful one can be measured by the difference between Martin Selge, *Adalbert Stifter: Poesie aus dem Geist der Naturwissenschaft* (Stuttgart: W. Kohlhammer, 1976), who operates with a general concept of "natural science," and recent studies such as Michael Gamper and Karl Wagner, eds., *Figuren der Übertragung*, and Christian van der Steeg, *Wissenskunst: Adalbert Stifter und Naturforscher auf Weltreise* (Zurich: Chronos, 2011), which explore specific scientific disciplines, discourses, and practices.

8 Geulen, "Tales of a Collector," 588.

9 In keeping with her fascinating monograph on Stifter, *Worthörig wider Willen*, Geulen favors a deconstructive resolution of the conflict between restoration and the "the notion of nature as an open-ended process," whereby "the enormous efforts required by this restoration project prove how deeply Stifter in fact registered the disintegration he sought to overcome or at least slow it down," "Tales of a Collector," 588. I merely argue that Stifter registers all this not only deeply, but also on the explicit surface of his texts.

strongly, for the words *Wiederherstellung* and reconstruction. What this normative use of the term "restoration" obscures, however, is that certain practices of reconstruction, including the one Stifter practices and that Viollet-Le-Duc choses to call "restauration," are inextricable from the discourse of geology that is central to the historical turn in nineteenth-century thought. Historical geology thus has an eminently modern message – deep time is a precondition of Darwinism – even as its practice is immanently reconstructive. Reconstruction, moreover, is not just a reaction to the destabilizing news of deep time. The imaginative reconstruction of a world before humans was a necessary step to establish the evidence of this world in the first place.

Stifter's participation in these modern reconstructive discourses in fact began before 1848, even if it did intensify when the Royal-Imperial Central Commission for the Study and Preservation of Architectural Monuments named him *Conservator* in 1853.[10] Geology appears in Stifter's early work in often surprising ways, or places. Nowhere is this more apparent than in the book *Wien und die Wiener in Bildern aus ihrem Leben* (1844: *Vienna and the Viennese in Images from Their Life*). Stifter edited this collection of non-fiction essays or tableaux (*Bilder*) of city life, and wrote several contributions himself. *Vienna and the Viennese* belongs to the genre of city "physiology," famously rediscovered by Walter Benjamin, who described it as "panoramic literature."[11] Such depictions of urban life were particularly popular on the French book market at the time, but also all across Europe, a fact that motivated the young author Stifter and his publisher Heckenast.[12] By invoking both *Bild* [image] and *Leben* [life] in the title, the work places itself in this genre tradition, which the French signal most frequently with *tableau* – the foundational text is Louis-Sébastien Mercier's *Le Tableau de Paris* (1781–88) – or indeed *physiologie.*

Stifter's preface to *Vienna and the Viennese* outlines the genre's poetic intentions by distinguishing between a quantitative survey and qualitative depiction of urban life:

> Es ist Zweck und Ziel dieser Blätter, nicht etwa eine *Statistik* Wiens zu bringen, sondern in ernsten und heitern Bildern, wie in einem Kaleidoskop, Scenen dieser Hauptstadt vorüber zu führen, die nur *ihr* zukommen, so daß sich dem Leser nach und nach ein Bild des Lebens und Treibens dieser Residenz zusammen male.

---

10 Cf. Wilfried Lipp, "Adalbert Stifter als 'Conservator,'" 186 f.

11 Walter Benjamin, "The Paris of the Second Empire in Baudelaire," in *The Writer of Modern Life: Essays on Charles Baudelaire*, ed. Michael Jennings (Harvard UP, 2006), 66.

12 On the French genre, cf. Richard Sieburth, "Une idéologie du lisible: Le phénomène des Physiologies," *Romantisme* 47 (1985): 39–60.

> The purpose of this collection is not to provide *statistics* about Vienna, but rather to pass in review, in images both serious and cheerful, as in a kaleidoscope, scenes of this capital city that belong to *her alone*, so that readers can paint for themselves, little by little, an image of the life and goings-on of this imperial residence.[13]

The ultimate goal of Stifter's book is to provide a unified or singular image (*Bild*) of Vienna, but this goal proves to be illusory, as it was, indeed, for Mercier in his *Tableau de Paris*, who had intended his portrait of the *grande ville* to serve as "l'abrégé de l'univers" [summary of the universe].[14] As Stifter's metaphor of the kaleidoscope – or the title page's image of scenes emerging from a fireworks spectacle (!) – demonstrates, city life contains a multitude of varied and transient images that cannot be subsumed in a handy summary. Indeed, unlike the panorama or the diorama, the mass media spectacles still alluded to in the verb "to pass review" (*vorüber führen*), the kaleidoscope *multiplies* rather than unifies images, even as its etymology ("seeing beautiful forms") aptly suggests the effects of poetic "glorification" (*Verklärung*).[15]

Yet what makes the modern metropolis most remarkable is also what makes it difficult to describe: its unprecedented size. With his characteristic flair for rhetorical antinomy, Stifter promises "einzelne Bilder und wieder auch Massen, das Gefühl des Einsamen und das Jubeln und Brausen der Menge" [individual images and once again masses, the feeling of the lonely man and the booming exultations of the crowd].[16] To judge from the topics Stifter himself chooses to address, however, he appears much more interested in the "masses" or "crowd" than in the individual.[17] In addition to the introductory panorama, "Aussicht und Betrachtungen von der Spitze des St. Stephansthurmes" [View and Observations from the Top of St. Stephan's Cathedral], Stifter's essays address a variety of mass urban phenomena from entertainment (the Prater park,

---

13 HKG 9.1, III.

14 Louis-Sébastien Mercier, *Tableau de Paris. Nouvelle édition* [1782], reprint, vol. 1 (Geneva: Slatkine, 1979), vi.

15 On the history of the panorama as a mass medium, see Stephan Oettermann, *The Panorama: History of a Mass Medium*, trans. Deborah Lucas Schneider (New York: Zone Books, 1997). Vance Byrd identifies three modes of representation in Stifter's book, "the panoramic, daguerreotypic, and kaleidoscopic," the last of which best corresponds to the work as a whole and to the "vitality and chaos of urban life." Cf. Vance Byrd, "Beautiful Form? Vienna and the Viennese and Stifter's Urban Sketches," *Interférences Littéraires/Literaire Interferenties* 8 (May 2012): 155–169.

16 HKG 9.1, 4.

17 For a comprehensive study of the topic of "masses," see Michael Gamper, *Masse lesen, Masse schreiben: eine Diskurs- und Imaginationsgeschichte der Menschenmenge 1765-1930* (Munich: Fink, 2007), and in particular Chapter 4, "Mittelmaß und Durchschnitt," 305–354.

Viennese salon scenes, country excursions) and festivals (Christmas, Holy Week) to advertising and the (flea) market, and even the postal system, as well as the final destination of all masses: the catacombs. Other topics, like meteorology ("Viennese Weather"), may not seem to involve the problematic of the masses at first glance, yet as we shall see, they demonstrate a key trope in Stifter's work: how the accumulation of small things can itself be viewed as great. Even an essay like "Die Streichmacher" [the tricksters], which appears to qualify as a portrait or description of moral types (*Sittengemälde*) that is a hallmark of urban physiologies, is an exception that proves the rule: Stifter is interested in the city as a mass phenomenon.[18] Even the tricks or cons of the "Streichmacher," or, to use Herman Melville's nearly contemporaneous phrase, the "confidence men," depend on the anonymity of the urban masses. Themselves "sehr zahlreich und fast so weit verbreitet, als das Gras" [quite numerous, nearly as plentiful as grass], the tricksters "sind immer dort am dichtesten, wo überhaupt die Menschen dichter wohnen, daher die meisten in großen Städten und an Orten, wo man sich drängt" [are most frequently found in places where people live in high density, thus in large cities and wherever people crowd together].[19] In other words, before Stifter became the author of the "little" people, of children, tar burners, cobblers and country priests in *Many-Colored Stones*, he was an author of the masses in *Vienna and the Viennese*.

The Urban Panorama: The View(s) from St. Stephen's Cathedral

If the kaleidoscope of the preface suggests one mode of apprehending the diversity of urban life, "View and Observations from the Top of St. Stephen's Cathedral" of the introductory chapter returns to the mode of panorama. Following the conventions of contemporary travel guides, including the Baedeker series established in 1832, this chapter takes an imaginary visitor to the top of the tower

---

18 On the role of types in the "physiologies" from a comparative perspective, see Olaf Briese, "'Typen' oder 'Individuen'? Personennamen in literarischen Sittenstudien um 1840," *Interférences littéraires/Literaire interferenties* 8 (May 2012): 135–151. The essay "Leben und Haushalt dreier Wiener Studenten" [The Household of Three Viennese Students] combines reflections on the student "type" with autobiographical conent.

19 HKG 9.1, 188; 189.

to provide an overview and survey of the city below.[20] Indeed, the church tower serves as a point of orientation from below as well, as a central landmark for the traveler navigating the winding city streets. In both instances, however, Stifter's text complicates the panorama's promise of orientation by emphasizing the bewildering size of the metropolis. The image of the city as a complex labyrinth corresponds to the cultural-critical tone that Stifter adopts throughout the volume. Sabine Schneider aptly calls Stifter's pessimistic diagnosis of modernity "cultural erosions."[21] As the phrase suggests, Schneider argues that order and disorder appear in Stifter's essay on both sides of the nature/culture divide, even as Stifter ostensibly turns to natural metaphors to critique and overcome cultural decline. If the panorama suggests a view of the city as a landscape, as a surveyable and transparent topography, then the geological truth of this landscape is its chaotic and unfathomable history.

In this reading, the panorama functions in much the same way as in a passage from the novel *Indian Summer.* Here, the view from above evokes "centralized order" as it visually confirms the laws of geological formation.[22] The high point provides both maximal visibility and represents the center around which the surrounding formations are organized. The young geologist Heinrich Drendorf compares the view of ice crystals in a microscope to the surface of the earth, knowing that in both cases, a (potentially) transparent process has produced the beautiful results:

> Alle diese Dinge stellen sich zu einem Ganzen zusammen, und die Strahlen die Thäler die Rücken die Knoten des Eises sind durch ein Vergrößerungsglas angesehen bewunderungswürdig. Eben so stellt sich von sehr hohen Bergen aus gesehen die niedriger liegende Gestaltung der Erde dar. Sie muß aus einem erstarrenden Stoffe entstanden sein, und streckt ihre Fächer und Palmen in großartigem Maßstabe aus. Der Berg selber, auf dem ich stehe, ist der weiße helle und sehr glänzende Punkt, den wir in der Mitte der zarten Gewebe unserer gefrorenen Fenster sehen.

> All these things come together as a whole, and the rays, valleys, ridges, and knots of ice are wondrous to behold when examined through a magnifying glass. In a like manner you can see the configurations of the Earth at the lesser altitudes from the peaks of very high mountains. The Earth must have originated from a hardening substance and expanded its parts to a tremendous magnitude. The mountain itself on which I was standing is the brilliantly gleaming white point which we see in the tender web of our frosted window.[23]

20 Significantly, the first Baedeker devoted to a large metropolis was the Paris guide from 1855, cf. Susanne Müller, *Die Welt des Baedeker: Eine Medienkulturgeschichte des Reiseführers 1830-1945* (Frankfurt a.M.: Campus, 2012), 193.

21 Schneider, "Kulturerosionen," 249–269.

22 Ibid. 256.

23 HKG 4.1, 43 [28].

As we shall see in the next chapter, Drendorf will later acknowledge how geological forces complicate this panoramic view of earth history, modify and confuse the "tender web," and thus obscure the presumed clarity of the processes involved in such change. Still, as Schneider suggests, panoramic descriptions in Stifter's work function first of all to establish representational order.

In his descriptions of the view from the tower of Saint Stephen, for example, Stifter compares both the tower itself and the city below to mountain formations. Stifter also wrote two versions of this essay, a fact that underscores its significance. In both versions, geological metaphors play an important role. The second essay, "Vom Sanct-Stephansthurme" [From Saint Stephen's Tower], was written ca. 1860 as part of an unfinished project to update *Vienna and the Viennese.* Here Stifter writes that the church is "wie ein Gebirge einfach und doch so schön" [like a mountain, so simple and yet so beautiful].[24] His motivation for writing this second version of the essay is not incidental to the problem of apprehending and describing the modern metropolis. As the intended preface for the project *Aus Wien* [from Vienna] suggests, Stifter returned to his Vienna writings precisely because of the extensive modernization that had changed the city in the intervening decades. Like his contemporary Baudelaire, Stifter saw his former home become historical: "Was ich hier von Wien sage, stammt aus Wien vor der so genannten Neugestaltung, also, wie sie jezt sagen, aus dem alten Wien" [What I say here about Vienna, comes from the Vienna before the so-called redesign, that is, as people say today, from old Vienna].[25] In other words, one can understand the later project *From Vienna* as a retrospective restoration of the book *Vienna and the Viennese.* Written after the novel *Indian Summer,* the second essay thus reflects important developments in Stifter's style, attitude towards the past, and knowledge of geology.

In this second version, it is significant that Stifter emphasizes the topical association of the panorama with the *mountain* panorama. While the first essay does naturalize the church tower, calling it a poplar (*Pappel*), it does not explicitly compare the tower to a mountain. Only in one section of the first essay – as the dawn illuminates the visitor's first view from the tower – does the description of the landscapes and mountains (e.g. Schneeberg) surrounding Vienna imply that the location of the observer is itself mountain-like in its height. Not surprisingly, the second version of the essay reprises this section, although it excludes

---

24 Adalbert Stifter, *Sämmtliche Werke*, vol. 15, ed. August Sauer (Prague: J.G. Calve, 1935), 380. Hereafter with the abbreviation PRA.

25 PRA 15, 376.

all the individual close-ups of city life that so inspired the younger Stifter. These close-up views in the first version more resemble what one might see from a "corner-window" than a church tower, and indeed, Stifter's earlier text has much in common with E.T.A. Hoffmann's proto-realistic *Des Vetters Eckfenster* (*The Cousin's Corner-Window*, 1822)[26] Only twenty years later, in the 'post-estival' revision, does Stifter emphasize the distance between tower and city below with mountain metaphor and its aesthetics of the sublime.

As opposed to a privileged view of the myriad, kaleidoscope images, the mountain metaphor in the later essay allows for the apprehension of the city as a whole via a dialectic of scale. First, Stifter describes how a traveler approaching the city will first be impressed by the tower's size from afar. Soon, however, the newcomer habituates to the impression of great size, and the city appears small when viewed from its nearest outskirts. Similarly, a mountain looks smallest when one perceives it from its foot. Only the ascent of mountain or tower can restore the impression of great size, as now other objects below can serve as points of comparison. Borrowed from the discourse of the sublime, this dialectic simplifies the impression of the city by reducing it to a question of scale.

In the earlier version of the essay, Stifter describes the city in geological terms that accentuate rather than reduce its complexity:

> ein Gewimmel und Geschiebe von Dächern, Giebeln, Schornsteinen, Thürmen, ein Durcheinanderhegen von Prismen, Würfeln, Piramiden, Parallelopipeden, Kuppeln, als sei das alles in toller Kristallisation an einander geschossen, und starre nun da so fort.
>
> a throng and heap of rooftops, gables, chimneys, towers, and intermixing of prisms, cubes, pyramids, parallelograms, cupulas, as if everything had been smashed together in wild crystallization and now remains frozen.[27]

This sentence contains in fact two geological concepts: the heap of rubble (*Geschiebe*) and crystallization. The description of this crystallization as "wild" or "mad," however, suggests that Stifter does not (yet) distinguish between the ordered production of crystals and chaotic destruction that leaves behind rubble.[28]

---

26 Hoffmann's short tale attracted Walter Benjamin's attention as "one of the earliest attempts to capture the street scene of a large city," Walter Benjamin, "On Some Motifs in Baudelaire," in *The Writer of Modern Life*, 189.

27 HKG 9.1, IX.

28 As the author will, for example, in the novel *Indian Summer*. Sabine Schneider's otherwise convincing essay, "Kulturerosionen," tends to ignore the development of Stifter's engagement in geology. If Stifter's style, as most readers agree, is characterized by a conflict between order and disorder, then the geology discussed in *Indian Summer* not only universalizes and naturalizes

Indeed, the geological metaphors here join other vivid similes: the city is "[w]ie eine ungeheure Wabe von Bienen" [like a monstrous beehive] and the streets like "Adern" [arteries].[29] Most prominent are oceanic metaphors, which are as commonplace in the nineteenth century as they are powerful. The city population floats "wie Treibholz, demselben Strudel zu" [like driftwood, towards the same maelstrom]. It is a "Strom" [current] that rushes until the "Wogen" [waves] flood St. Stephen's Square.[30] Even the milk these masses drink daily arrives in such great volumes that it could fill a "Weltmeer" [ocean].[31]

As if following Charles Lyell's call to portray rock too, and not only the sea as the "image of inconstancy," Stifter's text speaks of a "Steinmeer" [ocean of stone], conjuring the earth as a force as massive, mobile, and dynamic as the sea.[32] The essay invokes, in particular, the "Hereinschieben" [pushing in] of the "Masse der Vorstädte" [mass of suburbs] – a "mass" to which the city owes its "greatness" – against the former Glacis (itself a pile of earth!) of the fortified inner city.[33] The geological, glacial pushing (once again the semantics of *Geschiebe*) that underlies this military metaphor can cause, in another passage, a "Schutt von Häusern" [rubble of houses] hardly distinguishable from the "Staub" [dust] to which, so the narrator speculates (optimistically), even the prison building may one day be reduced. The text describes the prison as "jenes neue Haus, das sich so kolossal und fast egiptisch ernst und schwer dem Sandplatze entlang zieht" [that new building that stretches along the sand court so colossally and almost Egyptian in its gravity]; here the cipher "sand" both underscores the Egyptian metaphor and prefigures the erosion of the building to "dust."[34]

Unlike the aquatic metaphors of flowing traffic, streaming urban masses, or flooding of squares, however, such geological figures of speech, be they burying or piling up, are only partly metaphorical. After all, the city is made of earth and stone, a fact that Stifter's text repeatedly emphasizes:

> Ein großer Teil [der Wagen], die du mit einem dunkelroten Stoffe beladen siehts, kömmt von jener Gegend, aus der du hinter dem Berge einzelne Rauchsäulen aufsteigen siehest, und bringt unablässig und unermüdlich jenes Materiale, woraus sich dieses riesige Häuser-

this conflict, but also submits it to analysis – i.e., crystallization produces order, erosion disorder – and thus finds the order in disorder, so to speak.

29 HKG 9.1, IX; VI.

30 HKG 9.1, VI.

31 HKG 9.1, XVI.

32 HKG 9.1, VIII. Charles Lyell, *Principles of Geology*, 156.

33 HKG 9.1, X.

34 HKG 9.1, XV.

gewimmel nach und nach erbaut hat: die *Ziegel* – und im Wienerberge liegen noch und harren unmeßbare Schichten von Ton, daß man noch ein Wien und noch eins und weiß Gott wie viele aneinander fortbauen könnte, bis der Berg erschöpft und eben, aber auch von der Stadt verschlungen wäre! Und sieht man so zu, wie sie sich sputen und treiben und wirken, so sollte man meinen, sie hätten auch nichts Anderes im Sinne.

Most [of the wagons] that you see loaded with dark red material come from that region where you see individual columns of smoke rising behind the mountains. These wagons are constantly and tirelessly delivering the material from which this gigantic throng of houses has been built over time: the *bricks* – and in the Vienna mountain there still remain immeasurable layers of clay, enough to build another Vienna and then another and God only knows how many, until the mountain is emptied out and level, as if it had been eaten up by the city! And if you watch how they hustle and bustle and work, then you could easily believe that this was their very intent.[35]

This is the image that the panoramic view from St. Stephen's Cathedral affords, at least in Stifter's imagination: the modern metropolis moves mountains. Urbanization and urban growth progressively displace earth from the quarry and brick factory to the streets and houses of the city. In order to have such geological effects, the urban builders take on geological qualities: "immeasurable layers of clay" that thousands or millions of years of earth history had gradually deposited are here moved onward by the "unceasing" and "untiring" effort of modern industry.

Digging Deeper: "A Walk Through the Catacombs"

A variant of the threatening potential of both modern culture and geological forces appears in the second, almost equally famous essay of the collection as well: "Ein Gang durch die Katakomben" [A Walk Through the Catacombs]. If the "View and Observations from the Top of St. Stephen's Cathedral" presents a bewildering and at times overwhelming picture of the city, the "Walk Through the Catacombs" only amplifies this bewilderment through the inclusion of the city's depths, both spatial and temporal. One way to measure the extent of this confusion are the self-censoring elisions to the text that Stifter made in the later version, "From St. Stephen's Cathedral." For example, the second version replaces the anxious description of a tourist losing his way with the deadpan concession, "Zuweilen täuscht sich ein Neuangekommener" [sometimes a newcomer gets disoriented]. It also appends this brief summary of a visit to the catacombs:

35 HKG 9.1, XI.

> Einmal waren wir in den Katakomben, welche unter der Kirche und dem Thurme so wie unter Theilen des Stephansplazes und der Stadt mit unzähligen Wölbungen fortlaufen, und eine Menge geschichteter Menschenknochen und getrockneter Menschenleichen enthalten. Wieder an das Tageslicht getreten bestigen wir den Thurm, um den traurigen Eindruk von unten wieder in den offenen Lüften und in dem freien Leben hinweg zu bringen.
>
> Once we went down to the catacombs whose countless vaults extend from under the church and tower and parts of Stephen's square and the city, and which contain many layers of human bones and dried human corpses. Back in the light of day, we climbed the tower in order to dispel the sad impression from below in the open air and free life above.[36]

One can only speculate about whether Stifter planned to return to the catacombs with a full essay in *From Vienna*, but "From St. Stephen's Cathedral," which privileges the clear and centralized view of the panorama, shows little interest in getting lost among the winding streets or worse, in the dark catacombs below.

Yet this later, short description recalls key features of the earlier essay. First, the catacombs are a hidden realm of unanticipated and disconcerting size. As such, they are the cultural equivalent of the earth's own underworld, the countless subterranean layers that bear witness to a much longer history – the essay speaks of "Jahrmillionen" [millions of years] and "Billionen Jahrtausende" [trillions of millennia] – than the centuries suggested by the human remains.[37] From this geological perspective, of course, earth history contains human history, just as the earth contains its remains. So while the primary theme of Stifter's essay is a horrifying encounter with human death, which resembles other contemporary catacomb tales such as Edgar Allan Poe's *The Cask of Amontillado* (1846), it presents this theme within the context of geological change, the horizon of deep time, and even the idea of the death of the earth itself.

Poe's tale of murder by live burial in the catacombs ends with the murderer's ironic tribute to his victim, "*In pace requiescat!*"[38] For Stifter, the only thing worse than death is the disturbance of the dead's peace and quiet. At first, the essay appears to suggest that the main culprit in this regard is human history, the economic and social development that led to the establishment of the catacombs in the first place. As Stifter tells it, the cemetery once surrounded the cathedral and insulated it from the city, yet "industry" and "commerce" pushed the

36 PRA 15, 381.

37 HKG 9.1, 50.

38 Edgar Allan Poe, "The Cask of Amontillado," in *The Collected Works of Edgar Allan Poe. Vol. III: Tales and Sketches*, ed. Thomas Ollive Mabbott (Cambridge, MA: Harvard UP, 1978), 1266. For a rare comparison of the two authors, cf. Sonja Osterwalder, "Trostlose Wissenschaft, schreckliches Leben. Adalbert Stifter und Edgar Allan Poe," in *Figuren der Übertragung*, ed. Gamper and Wagner, 273–288.

monuments to the dead back to the walls of the church.[39] To create a paved open square around the cathedral, the grave hills were leveled, and the bones of their inhabitants were revealed and displaced, desecrated and devalued, “jetzt legt man eure Reste, die Niemand kennt, wie das werthloseste Ding auf einen Haufen, um sie an einen anderen Ort zu bringen, wo sie wieder nicht bleiben” [now one throws your remains, which no one recognizes, into a pile, like the most worthless thing in the world, to be brought to another place, where they still will not rest].[40]

Displacement or movement and “leveling” (*Planierung*) are part of a critical semantic field that associates smoothness with circulation, progress, superficiality, and “Indifferentismus” [indifferentism].[41] In Stifter’s text, indifferentism appears to go beyond questions of religious doctrine. It is shorthand for the general “Glätte und Verflachung unserer Zeit” [smoothness and flatness of our times], for the interchangeability of words, ideas, and values.[42] Stifter illustrates modern indifference with the image of eroding pebbles, “durch vervielfältigte geistige und leibliche Communicationsmittel sind wir feiner, glatter, geschmeidiger geworden, wie Kiesel, die sich an einander abreiben: aber ist deshalb der Kiesel innerlich weniger hart?” [the multiplication of spiritual and material means of communication have made us finer, smoother and sleeker, like pebbles that rub themselves smooth: but does this make the inside of the pebble any less hard?].[43] Intensified communication and circulation polish people into “pebbles.” And it is fine pebbles and paving stones that produce the smooth surface on which the carriages roll throughout the essay. Such rolling functions as an acoustic reminder of modern life bustling above while the narrator visits the quiet catacombs below. As for the “spiritual” media of communication, Stifter does not fail to provide here a commonplace critique of excessive book printing, “daß, seit sie die Bücher so vervielfältigen, tausend und tausend Menschen aus der Welt gehen, ohne darin einen einzigen Gedanken gehabt zu haben” [that,

---

39 HKG 9.1, 51. Stifter’s discussion of the changes in the cemetery resembles the debate in Goethe’s *Wahlverwandschaften* about the same issue, see David Wellbery, “Die Wahlverwandtschaften,” in: *Goethes Erzählwerk: Interpretationen*, ed. Paul Michael Lützeler (Stuttgart: Reclam, 1985), 291–318; and Eva Horn, *Trauer Schreiben: Die Toten im Text der Goethezeit* (Munich: Fink, 1998).

40 HKG 9.1, 52.

41 Stifter may have learned the term *Indifferentismus* from Pope Gregor XVI’s first encyclical *Mirare vos* (1832), though while he uses the term in its religious sense of “indifference to questions of religious faith,” his essay ultimately “secularizes” this usage by giving it myriad material forms (level roads, heaps of bones etc.).

42 HKG 9.1, 50.

43 HKG 9.1, 49.

ever since they have printed books in such numbers, thousands and thousands of people leave the world behind without having thought a single thought].[44]

Though the first-person narrator visits the catacombs himself, he belittles the motivations of the other visitors to the catacombs. Driven by "Neugierde" [curiosity], or seeking the "Schauergefühl" [frisson] of an encounter with death, the typical visitor is marked by "frevle[m] Vorwitz" [blasphemous impudence].[45] For his part, the narrator hopes piously to give the readers "ein ernstes Stück Vergangenheit" [a serious dose of the past] so that he might "etwa anfange, über Gott, über Weltgeschichte, Ewigkeit, Vergeltung u.s.w. nachzudenken, und vielleicht ein anderer werden" [begin to think about God, about world history, eternity, reckoning and so forth, and perhaps become another person].[46] Yet instead, the narrator finds nothing in the catacombs but corpses, "starr, über einander geschichtet, eine werthlose, schauererregende Masse" [stiff, in layers, a worthless, horrifying mass].[47] The underground passages are scattered with "Trümmer" [ruins], the walls constructed of bones: "Ein Führer bedeutete uns, daß man die vielfach zerstreuten Knochen der Katakomben, und die einst auf dem Stephansfriedhofe ausgegrabenen hier der Ordnung wegen aufgeschichtet habe" [A guide told us that that the bones, once scattered across Stephen's cemetery, had been dug up, and piled up here in layers to create some order].[48] Not only neglect and decay, but also renovation and order lead to the ultimate confusion.[49] As on the roads above, paved by smooth, indifferent stones, the paths below are made of the dust of long-decayed bodies, their softness the only remaining hint of the organic. In this confusion, it is hardly a surprise that the narrator fears losing his way, being left behind by the group to die alone, and hence hangs on to every glimmer of light, to every sound of life above, be it carriages or church music, as a sign that he is not completely forsaken.

Unexpectedly, the air down in the catacombs is fresh: "Die Luft trug nichts bei; denn trotz den hier vorgegangenen Acten der Zersetzung waren diese doch schon vor so vieler Zeit, und es ist seitdem eine solche Trockenheit eingetreten, daß die Luft, durch viele Schachte in Communication mit der äußern erhalten, ganz trocken und rein ist" [The air did nothing to add to the effect; despite

44 HKG 9.1, 50.
45 HKG 9.1, 53.
46 HKG 9.1, 51.
47 HKG 9.1, 54.
48 HKG 9.1, 54.
49 This thought applies to all renovations and restorations, including those Stifter depicts more positively, cf. Saskia Haag, "Versetzt. Restaurierung als Entortung in Stifters Nachsommer," in *Figuren der Übertragung*, ed. Gamper and Wagner, 77–87.

the processes of decomposition that had taken place here long ago, by now there is such a dryness that the air, which communicates through many shafts with the outer world, remains dry and pure].[50] Foul air is a common feature of the underground, in particular in its horrific variants. Poe speaks of "a deep crypt, in which the foulness of the air caused our flambeaux rather to glow than flame."[51] In his novella *Turmalin*, Stifter employs the unhealthy air of a cellar apartment as a central motif. Is the fresh air in the "Catacombs"-essay, which otherwise hardly spares in the rhetoric of shock and horror, merely a concession to the demands of realism? This seems unlikely, given the significant poetic license that Stifter takes in terms of the location, depth and breadth of the catacombs.[52]

The "communication" of the air communicates that the underground world is not so separate from the world above after all. Against the backdrop of traditional knowledge that separates our world from the underworld as hell, Hades etc., Stifter's essay presents geological knowledge that refuses to recognize a difference between these realms. In this respect, the essay's content contradicts its traditional narrative structure, which recounts the descent with a guide, the encounter with the dead, and a perilous and uncertain return to the world of the living. Despite some hedging about causes – at one point the narrator notes "aus einer mir unbekannten Ursache war hier keine Verwesung eingetreten" [for some reason unknown to me no decay had set in here] – his comments about decay and decomposition make clear that he understands them as natural processes that can be studied scientifically.[53] For example, the narrator speculates that "In diese verschlossene Räume muß die Verwesung nicht haben eindringen können, so daß die organischen Gebilde blos vertrockneten, aber nicht zerstört wurden" [these closed rooms must have kept the decay out such that the organic forms only dried out, but were not destroyed].[54] Dried but not destroyed, these "organic forms" resemble fossils. Remarkably preserved, they are what brings the past into the present – or the present back to the past, as we shall see below.

That the air, and the forces of nature more generally, must in the end lead to disruption, displacement and destruction, however, is the moral of a "legend"

---

50 HKG 9.1, 54.

51 Poe, "The Cask of Amontillado," 1261.

52 Such license, along with deviations from literary predecessors, is the main object of Gustav Wilhelm's commentary in the Prague-Reichenberger edition of Stifter's works, cf. PRA 15, 401–403.

53 HKG 9.1, 56.

54 HKG 9.1, 55.

that demonstrates how even the best efforts to protect the eternal peace of the dead are in vain. After his death, Attila the Hun was placed in a golden coffin, which itself was enclosed in a silver one, then in iron, and finally, in stone. Then the gravediggers diverted a river, dug a deep grave into the bed, and let the water flow above it. As a reward for their work, the gravediggers were then killed, so that no one would ever find out where the King was buried:

> Aber eines Tages wird der Fluß den Sand und Schlamm in einer Ueberschwemmung herausstoßen, oder man wird eine Wasserbaute anlegen, oder der Fluß wird seinen Lauf ändern, und man wird im alten Bette ein Feld oder einen Garten graben: dieses Tages wird man dann den Sarg finden, das Gold und Silber nehmen, den König aber hinaus werfen auf den Anger der Haide.

> But one day the river will push out the sand and mud in a flood, or somebody will build a dam, or the river will change its course, and people will dig out a field or a garden in the old bed: on this day someone will find the coffin, take the gold and silver, and throw the king out onto the meadow.[55]

A riverbed may seem like a strange place to hide a coffin, but the message of the legend is clear: every "bed" – perhaps even the "bedrock" – is like this riverbed, subject to flow and change.

There is literally "no place" on earth, or indeed in "this universe" that is stable enough to be called an eternal resting place. While the geology of "A Walk Through the Catacombs" may be imprecise, the impact of the geological outrage to human narcissism is clearer here than at any other place in Stifter's work: "Als ob irgend auf der Erde etwas ungestörtes, etwas unvergängliches wäre! Ja, ist nicht am Ende sie selber vergänglich, und wird eine Leiche, so wie die, die man jetzt so sorglich in ihrem Bauche verbirgt?" [As if there were anything on earth undisturbed or intransient! Indeed, is the earth not in the end itself mortal, and will it not become a corpse just like the ones that we hide so carefully inside its belly?].[56] To call this scenario apocalyptic is almost an understatement: the idea that the earth will one day become a corpse brings the terrestrial apocalypse down to the same level as the daily occurrence of human death.

As Stifter's essay goes on to demonstrate, the so-called geological outrage is a complement to the "cosmological outrage" (*Kränkung*) that Freud includes in his famous list.[57] Viewed from the Sirius star by an imaginary extraterrestrial, the

---

55 HKG 9.1, 58.

56 HKG 9.1, 58.

57 Freud, "Eine Schwierigkeit der Psychoanalyse," 3 – 12. Indeed, when nineteenth-century authors return to the shock of Copernicanism and write exclusively about astronomical dimensions, it may be geology that gives the shock its energy.

death of the earth would pass unobserved because it would be invisible, "hätten sie dort auch Röhre, die zehntausendmal mehr vergrößerten als die unsren" [even if they had telescopes that could magnify ten thousand times more than our own].[58] Earth and its inhabitants are too minuscule to be visible on a cosmic scale. To return to Friedrich Nietzsche, as quoted in the introduction, the life of the earth itself becomes as important as the "Weltgeschichte der Mücken oder der Eintagsfliegen" [world history of mosquitoes and mayflies].[59]

Can this be the meaning of "God, world history, Eternity etc.," the search for which motivated the descent into the catacombs in the first place? The question is whether the "world history of mosquitoes or mayflies" is the same as the history nestled between God and Eternity. At the very least, the feelings associated with each form of contemplation could not be more different. In the context of a theological eschatology, the reminder of one's own mortality or the world's temporality may be somber, but is ultimately uplifted by thoughts of immortality and eternity. Hence the cemetery and catacomb are sites of rest, calm, and refuge from the turbulence of city life. As soon as one's life is set against the long duration of geological and cosmic time, however, the consoling horizon of eternity disappears. All that is not permanent now becomes a source of trouble and irritation. As we saw above, transience (*Vergänglichkeit*) and disruption (*Störung*) are synonymous and equally omnipresent, "As if there were anything on earth undisturbed or intransient!" The qualitative difference between time and eternity becomes a quantitative difference between short and long times. No longer a placeholder in the here and now pointing to the eternal rest beyond, the grave is as subject to change as any other place on earth.

The Catacombs Restored

These images of constant and unstoppable erosion may elicit from the narrator panic or pessimism, but they also motivate fantasies of reconstruction and restoration. At the start of "A Walk Through the Catacombs," the narrator imagines how a restored churchyard would provide the proper setting for the earnest reflection he desires:

> wie wäre es religiös feierlich, wenn jede Kirche, selbst in den Städten, mit einem großen Garten der Todten umgeben wäre, der durch eine Mauer von der leichten Lust der Lebenden

---

58 HKG 9.1, 58.

59 Nietzsche, "Über Wahrheit und Lüge im außermoralischen Sinne: § 1," http://www.nietzschesource.org/#eKGWB/WL-1; "Truth and Lying," 42.

> getrennt wäre, daß sie ein Gedanke der Ewigkeit anwandeln müßte, wenn sie durch das Gitter einträten.
>
> how religious and solemn it would be if every church, even in the big cities, was surrounded by a grand garden of the dead, separated from the frivolous pleasures of the living by a wall, such that a thought of eternity might come over them when they enter through the gates.[60]

The series of verbs in the *irrealis* mood ("wäre"/"would be," etc.) underscores not just how unrealistic, but also how highly desirable this restoration of the churchyard – or more precisely: of all churchyards (!) – would be. Yet the reasoning here also suggests how the fantasy is not so much one of re-sacralization as it is of musealization.[61] As much as one might consider the "thought of eternity" sacred, the text leaves out the theological idea of eternal life in favor of a secularized aesthetic of the *memento mori*, that is, the horrifyingly sublime thought of eternal death.

Musealization can achieve other aesthetic effects, as well. The modern culture that the essayist accuses of indifference to the past proves more concerned with preserving memory than he would like to admit. The narrator does not venture into the catacombs alone, after all, but takes part in a *guided tour.* Early in the tour, as some of the guests examine a skull, the guide comments, "Lassen Sie das liegen [...] wir werden schon noch mehr solches und besser erhalten antreffen" [leave that be [...] we will see more like it and better preserved at that].[62] Of course, they do encounter "better preserved" corpses, in arrangements and poses that appear staged. In one room "stand ein Sarg, ein einziger in dieser Gewölbe, als wäre er von seinem Ort absichtlich hieher gebracht, und geöffnet worden, und dann stehen gelassen" [stands a coffin, singular in this vault, as if it had been brought here intentionally from its resting place, and opened, and then left standing up].[63] In another, bodies lean against walls or crouch, all with folded hands, "aber bis zum Erschrecken deutlich waren die Gesichter und die Körperformen, als wären sie gestern hieher gestellt worden" [the faces and body parts were so well-defined it was horrifying, as if they had been placed here just yesterday].[64] Stifter's hesitation to acknowledge these arrangements ex-

60 HKG 9.1, 51.
61 Katharina Grätz discusses the catacomb essay at a key moment in her argument about musealization. The decay and morbid curiosity motivate Stifter to present museum preservation as a solution to the modern disregard of historical values. She does not address the extent to which the catacombs have already become a museum, cf. *Musealer Historismus*, 101–104.
62 HKG 9.1, 54.
63 HKG 9.1, 54.
64 HKG 9.1, 56.

plicitly as staged is not surprising and may be programmatic. After all, the text itself has the same intention as the guided tour: to lead readers into the catacombs so that they may reflect with the proper pathos on death and eternity, not indulge in the kitsch and bad taste of a morbid spectacle. Yet the cultural preservation at work is undeniable. As the tour guide explains, future archeological digs are already planned: "Es wird hier, wenn einmal alles ausgegraben und gelüftet sein wird, noch viel weitläufiger und wunderbarer herum zu gehen sein, als jetzt" [Once everything is dug and aired out, it will be much more spacious and wonderful to walk around here than it is at the moment].[65] The catacombs will become smooth like the streets above, but in the name of conservation of the past, not in the name of progress.

Given the essay's repeated and emphatic complaints that the dead can find no rest, moreover, the way the narrator describes how he and others poke the corpses with their canes seems perverse. Underlying this disturbing detail, however, is an important question: How should one act with respect to the material remains of the past? Nature appears indifferent, or worse, nature acts constantly to destroy what it has created. If there is an insight from the descent into the catacombs, it is that culture, a cult of the dead, erects monuments to designate this loss. These monuments, in turn, are subject to degradation and destruction. The "Walk Through the Catacombs" thus describes an experience that closely parallels Heinrich Drendorf's encounter with a heap of rubble in *Indian Summer*, discussed in the next chapter. The essay also gives expression to a complex set of feelings about the past: a horror of decay and rubble combined with awe and solemn respect for monuments or ruins. Beginning with the peaceful fantasy of a restored graveyard in front of the cathedral, the essay (unintentionally) describes instead the real, ongoing restoration of the catacombs themselves.

Recovering Value at the Flea Market

At the start of his essay on "Der Tandelmarkt" [The Flea Market], a trove of the younger Stifter's thoughts on the conservation of antiquities, the author refers explicitly to the competing roles of words and things in the telling of history:

> Der Referent dieser Zeilen gesteht, daß er ein großer Verehrer von Alterthümern ist, gleichsam von Worten, die eine längstvorübergegangene Zeit an die unsere redet, oder vielmehr: die Worte lasen wir in irgend einer Geschichte, und die Alterthümer sind der sinnlich eindringende Commentar dazu.

---

65 HKG 9.1, 60.

> The writer of these lines admits that he is a great admirer of antiquities, which are like words that a long-past age speaks to our own, or rather: we read the words in some history, and the antiquities are its sensuous, immersive commentary.[66]

Both these analogies insist on the priority of language in historical discourse. In the first, simpler analogy, the subordination of antiquities to words is merely implicit: words are essentially communicative; antiquities are only "like words" to the extent that they speak to us. The second analogy employs the philological hierarchy of text over commentary to make the priority more explicit. We must know something about history such that the things of the past have any meaning for us, and this knowledge is obviously transmitted via words. Yet what kind of commentary is "sensuous, immersive" ("eindringend" literally means "penetrating")? The words, which seem to have a logical priority over things, are themselves forgettable. One reads them in "irgend einer Geschichte" [some history or another]. But the thing has a "penetrating" aesthetic quality, which in the first sentence is not only "sensuous," but also "moving" and "naïve," qualities that likewise upend the priority of language and text. As media of communication, however, there is a problem with things. As fragments of the world to which they belong, they can only tell a fragmentary story. Stifter gives the example of "das Gewand, das der Urgroßvater ausgezogen und niedergelegt hat, als er auf immer fortgieng, welches Gewand nun rührend und naiv Bruchstücke der Geschichte dieses Urgroßvaters erzählt" [the clothing that the great-grandfather took off and set down, right before he left the world for good. Now this clothing tells us movingly and naïvely fragments of the story of this great-grandfather].[67] Material things are always only fragments of a story or history. They do not so much resemble commentary, as they require commentary.

"The Flea Market" has a similar structure to the "Catacomb" essay. It begins with a complaint against the indifference of modern culture to history and tradition, and ends with the same uncomfortable recognition that the antiquarian spirit may be more widespread than initially assumed. The problem at the start is the same, and the examples are only slightly different. Where the essay on the catacombs focuses on the level and smooth streets, Stifter criticizes here the uniformity of modern architecture, "ein ordentliches, vernünftiges Haus aber heißen sie ein großes gerades viereckiges Ding mit vielen Fenstern, das erst recht einem Kasten ähnlich sieht" [but they call a right, reasonable house a big, straight, rectangular thing with many windows, that looks like noth-

66 HKG 9.1, 227.
67 HKG 9.1, 227.

ing more than a box].[68] Just as the removal of gravestones destroys the indexical relationship of monuments to their significance, here street names lose their meaning when the streets change in architectural character and are renovated or destroyed. "Nur die Namen als Denkmale der Vergangenheit bleiben, aber auch die nicht immer" [only the names remain as monuments to the past, and not always even these], complains the essayist.[69] The erasure of history clearly does not bother with the finer distinctions between word and thing that a deconstruction of Stifter's text reveals.

Like the catacombs, the flea market is a place where one can find traces of forgotten history, but like death, the great equalizer of lives, the market makes all things equal: "Alle Stände, alle Alter und Geschlechter, alle Zeiten sind hier repräsentirt" [all classes, all ages and races, all periods are represented here].[70] The anthropomorphism – commodities as social classes – sustains the analogy from the catacombs between persons and things, and between liquidation and death. The flea market also functions as a kind of graveyard for things, "so haben die Sachen als letztes Ziel den Tandelmarkt, wo sie zum weiteren Vertriebe ausgestellt prangen" [thus things have as their final destination the flea market, where they are put up for sale once again].[71] This sentence of course contains a paradox, if not a direct contradiction: the flea market is both a final destination and a relay point for further commerce. However, the graveyard itself proved to be a temporary resting place, a reserve from which things may be moved again or fall to dust.

Because the fate of things in the market depends on their commercial value, the flea market also becomes a place of uncertain and hidden values. All of the anecdotes Stifter tells about the flea market have this in common: something is under- or overvalued; something reveals its true value only later, after the transaction is complete. Thus, a military jacket, sold by a grave-robber (!) for its value as clothing, in fact contains in its pocket a great sum of money that the old soldier had withdrawn from the bank shortly before his death. A clever shopper buys tin plates at a low price by convincing an old woman than his purchase is a charitable act. In another anecdote, a seller foists a number of old and blackened paintings on a woman who is interested in a number of other items. Thinking she can at least save the gilded frames, she takes them, though her unskilled attempt to clean the frames removes the gold leaf. When a friend convinces her

68 HKG 9.1, 227. "Erst recht" because "they" always say: "der alte unnütze Kasten [muß] weg," when it is time to destroy an old building.

69 HKG 9.1, 228.

70 HKG 9.1, 231.

71 HKG 9.1, 231.

to have the paintings themselves cleaned as well, two original Teniers appear behind the layers of soot. A fortuitous find.

The anecdote of the paintings is paradigmatic because it expresses the fantasy of every antiquarian plunderer who sorts through piles of junk at the flea market. The desire (*Gier*) of these searchers, the essayist included, is to uncover a secret treasure under the layers of dust and neglect. The anecdote illustrates that art history, like history more generally, is not a matter of constant, unchanging values. Just as styles change, so too the interest in a specific historical period. Historicism not only revives a general interest in the past, but also relativizes the worth of each past period by setting them against each other. One buyer might place a high price on Roman antiquities, while another cherishes Dutch genre painting. The essayist must admit, however, how rare any of these spectacular finds are, especially in proportion to the number of people looking for them.

The flea market thus reveals the strange economy of old things: to the extent that the mass of old objects becomes increasingly worthless, so too the value of the individual object, selected and restored, skyrockets. What is more, the seemingly irregular commerce of the antiques shopper becomes a predictable form of demand. Towards the end of the essay, Stifter offers the example of small cloth fragments called "Flecke" [patches]. He wonders whether an "industrious class" might be producing these scraps – presumably by cutting up larger, more useful swaths – because there is a market for them.[72]

The fact that trade in antiquities differs only superficially from the trade in new commodities irritates the essayist, who professes his love of antiquities and laments finding new goods at the flea market. The topos of modern indifference thus reappears here as the general equivalence of all commodities as goods of exchange. What, then, is the proper way to deal with old things? After the initial salvo against modern indifference to history, the essayist admits that objects, "die etwa historischen, künstlerischen, antiquarischen Werth haben, wandern in Sammlungen" [which have historical, artistic, antiquarian value, [do] find

---

72 Stifter is coquettish, fearing reproach for this comment: "ich fürchte, man mache mir den Vorwurf, ich satirisire auf die Sünde eines ganzen Standes, an deren Dasein ich doch eigentlich gar nicht glaube" [I fear someone might blame me for satirizing the sin of an entire class of people whom I do not even really believe exist], HKG 9.1, 240. The context here is unclear, although it may be a satire of the *Lumpensammlerin,* the rag-collector type characterized in a separate essay by Sylvester Wagner. It seems clear enough that Stifter believes that the merchants, who proffer such shreds of cloth, while they may be "cheating" the public, do not have an excessively profitable business.

their way into collections].[73] Yet collections fail to speak (to Stifter) as antiquities should, "like words that a long-past age speaks to our own," for the very reason that historians have selected them as so important:

> in solchen Sammlungen [sind] meist nur Dinge, die uns von dem vergangenen Staatsleben erzählen, aber gerade von dem nichts, was uns das unmittelbarste und herzigste ist, von dem alltäglichen Alltagsleben unserer Voreltern, von dem gerade der Plunder und Trödel deutlicher redet, als das wichtige Geschichtsstück, das am Ende doch wieder nur das Schluß- und Endestück des Trödels ist, und aus seinem Zusammenhange gerissen, verstummt [...].
>
> such collections hold for the most part only things that tell us something about political history, but when it comes to the nothing that is most immediate and heartfelt to us, about which the junk and rubbish speak more clearly than the important historical artefact is in the end just the tail and end piece of rummage (*Trödel*), and taken out of context (*Zusammenhang*), it falls silent.[74]

The polyptoton (*alltäglich, Alltagsleben*) marks a revaluation of values at the center of nineteenth-century historicism. The everyday cannot be quotidian enough. Like the flea market, the category of the everyday includes what every other category of value excludes, be it historical, artistic, or antiquarian. From the historical perspective that is only interested in the life of the state, everyday life is so trivial as to be "nichts" [nothing]. Yet this "nothing" becomes everything, or at least, it speaks more directly to the heart than "important" historical artefacts.[75]

Another word for this new perspective on history, which focuses on objects of everyday life, is "cultural history" or *Culturgeschichte*. One of the most prominent cultural historians of Stifter's epoch, Jacob Burckhardt, describes the shift from "state history" to "cultural history" in his lectures "On the Study of History" (*Über das Studium der Geschichte*), delivered at the University Basel from 1868 to 1872.[76] Burckhardt sees "cultural history," which he also calls the "universal perspective," as the key task and challenge of the contemporary moment:

---

73 HKG 9.1, 228. The concession demonstrates Stifter's awareness of the increasingly role of museum as a "Rettungsanstalt" [institution of saving, preservation], cf. Hermann Lübbe, *Der Fortschritt und das Museum. Über den Grund unseres Vergnügens an historischen Gegenständen* (London: Inst. of Germanic Studies, 1982), 14, and Katharina Grätz, "Musealisierung und Sammlung," in *Stifter-Handbuch*, ed. Begemann and Guiriato, 305. The latent critique of museums points to the alternative poetics of conservation that he will pursue in his literary works.

74 HKG 9.1, 228.

75 HKG 9.1, 228.

76 On the distinction between political and cultural history, cf. Felix Gilbert, *History: Politics or Culture. Reflections on Ranke and Burckhardt* (Princeton: Princeton UP, 1990). Katrin Maurer compares Stifter's descriptive style with that of the historian, cf. K. Maurer, "Close-Ups of Histo-

> Gegenwärtig: der universelle Standpunkt; das Sammeln der Facta nicht mehr allein nach gewissen äußerlichen Beziehungen, sondern als Karakteristik der Zeiten. Plötzliche Bedeutung zahlloser einzelner Data; neben die Staatsgeschichte etc. stellt sich eine endlos weite Culturgeschichte; [...] Alles Erhaltene wird zum redenden Zeugniß der betreffenden Epoche, zum Monument [...] Auf den geschichtlichen Gehalt aller Monumente hinzuweisen, wäre die Hauptaufgabe für uns.
>
> Presently: the universal perspective; to collect facts not only with respect to certain external relationships, but also as the characterization of historical periods. Sudden significance of countless individual data points; in addition to state history etc. there exists an endless field of cultural history; [...] Everything that has been conserved becomes the speaking witness of an epoch, a monument [...] Our primary task would be to identify the historical meaning of all monuments.[77]

Though Burckhardt's texts avoids the (ironically) pejorative terms "junk" and "rummage" (*Plunder, Trödel*) so dear to the young Stifter, it is clear from his use of the term *Fakta* and *Data*, that he means here not the "dates" (*Daten*) of political history, but the raw "facts and data" of material culture. The key term in Burckhardt's text is thus "everything;" an indefinite, "endless" totality that is limited only by the erosive effects of time. The Burckhardtian cultural historian, that is, works within the same limits as the archeologist or the geologist: "everything that has been conserved."

Stifter's earlier text also provides a critical commentary to what Burckhardt will call "certain external relationships." In the context of political history, the significance of a cultural object depends on its relationship to a political figure or event, for example: the dagger used to kill Julius Caesar. In the context of cultural history, every dagger has a significance, or, to offer an example from Stifter's essay, the Roman sword used as a theater prop would tell us something about the (recent) history of the theater, despite being a "fake" relic of Roman history. Indeed, as Stifter emphasizes with a rhetorical chiasm: rummage (*Trödel*) is *more* important than the "most important historical" artefact, which "in the end" is *less* important than "the tail and end piece of rummage" (*Trödel*). The appeal to the long run (here: *am Ende*) implies the same cosmic relativism of the "Catacomb" essay, the deep time of earth history that swallows even the most important objects of human history. By contrast, the *Zusammenhang* evokes another idea that will become central to Stifter's restorative program and practice. What matters most in this paradigm is not the absolute significance of an artefact, but the extent to which it reveals the milieu of its former use and func-

---

ry: Photographic Description in the Works of Jacob Burckhardt and Adalbert Stifter," *Monatshefte* 97, no. 1 (2005): 63–77.

77 Jacob Burckhardt, *Über das Studium der Geschichte*, ed. Peter Ganz (Munich: Beck, 1982), 84.

tion. The essay underscores this thought that "zweitens reden ja die Sachen am rührendsten zu uns von dem Ort ihres einstigen Gebrauchs aus und im Zusammenhang ihrer Umgebung; werden sie nun von da herausgerissen und in einen fremden Saal zusammengedrängt, so verlieren sie meistens ihre Muttersprache" [things speak to us most movingly from the place where they were once used and in the context of their environment; if they are ripped out of there and squeezed into a foreign museum hall, they most often lose their mother language].[78] As Stifter's linguistic metaphor suggests, this model has consequences for both things and words, for both material and literary-cultural restorations.

As Saskia Haag has demonstrated in her reading of *Indian Summer*, the notion that cultural authenticity is bound to place of origin is undercut by the practical effects of restoration.[79] Restoration work always dis-places even as it restores, even when the goal is to restore something to its original place. And as we learned in the catacombs, earth and human history are always displacing everything. As in the "Catacomb" essay, moreover, Stifter demonstrates in his reflections on antiquities at the flea market a remarkable sensitivity for the ambivalence of restoration and conservation efforts. After discussing the significance of antiques and how they "speak [...] most movingly in the context of their environment," Stifter relates an anecdote about a peculiar antiquarian, whom he calls a "great fool" (*großen Narr*). The term *Narr*, which reappears at the end of the essay and in the story "Castle of Fools" (*Die Narrenburg*, discussed at the end of the chapter), is itself ambivalent: while it is clearly pejorative, its use is almost exclusively ironic and self-deprecating. *Narr* is thus a mark of distinction.[80]

The distinguished "fool" in the anecdote inherits a country house that has not been occupied for hundreds of years, moves in, and does not change a thing. Since some things are inevitably in need of repair, this man "restores" them according to a principle that appears to contradict the commonsense view on such matters:

> Nur hatte er die Ansicht [...] daß man die Ausbesserungen nicht etwa im Geiste und Style des bereits Vorhandenen machen müsse, wie ich und die andern Vernünftigen gethan hätten, sondern wo etwa ein Fensterflügel fehlte, setzte er neben den alten einen ganz mod-

78 HKG 9.1, 228.

79 Haag, "Versetzt. Restaurierung als Entortung," 77–87.

80 Ulrich Dittmann comments that *Narr*, in addition to being fashionable at the time, was one of Stifter's "favorite words," and cites, for example, the end of "The Flea Market:" "einen Narren heißen sie mich – mag sein, ich beneide den nicht, der keine Narrheit hat" [They call me a fool – that may be, but I do not envy those who have no foolishness], cf. "Stellenkommentar," HKG 1.9, 129.

ernen hin; der alterthümliche invalide Sorgenstuhl bekam einen Fuß heutiger Art; wo das Wandgetäfel fehlte, ließ er die nackte Mauer durchblicken, nur ward das Stück betüncht und nach jetztiger Weise bemalt.

Now he believed [...] that improvements must not be done in the spirit and style of the existing house, as I and other reasonable people would have done, but rather that when a window casement is missing, he would put a brand new one in next to the old one; an antique, sickly, shaky chair would receive a new foot in today's style; if part of the wall paneling was missing, he would let the bare wall peek out, whitewashed and painted as one does today.[81]

The essay goes on to justify this approach by appealing to an aesthetics of contrast. One can more easily recognize the value of the old through its juxtaposition with the new: "gerade der Kontrast hob uns das urväterliche heraus, und da nirgends nachgeholfen war, so erwies sich das Alte als echt und wirkte als solches" [the contrast highlighted the venerability for us, and since no maintenance had been undertaken, the old parts remained authentic and appeared as such].[82] As the narrator reports, he comes around to this "foolish" view, as does, indeed, the discipline of conservation and restoration as it is practiced today. As we shall see in the next chapter, Stifter himself eventually embraces the view of "reasonable people," that one should restore "in the spirit and style" of what remains from the past.[83] By providing evidence of a diverging view, the flea market essay not only reveals the historical contingency of "restoration," but also how Stifter will continue to employ literary technique – be it the irony of the word "fool" or the contradiction between discourse and plot in *Indian Summer* – to explore the complexities and ambivalence of the historicist desire to reconstruct the past.

Cultural Sedimentation: *Limestone*

Stifter's essays in *Vienna and the Viennese* demonstrate anxiety about the changes brought on by rapid modernization, yet express this concern via natural metaphors, which, consciously or not, also convey the disturbing message of geological deep time. At various points, restoration appears as a possible solution or, at least, as a desirable response to the erosive dangers of nature and culture alike. The essays thus lay important groundwork for the paradigm of geohistoricism that the author will describe in detail, and with the benefits of self-

81 HKG 9.1, 229.

82 HKG 9.1, 229.

83 Without discussing "The Flea Market," Wilfried Lipp notes the fluidity of restoration concepts in Austria in the 1840s and 1850s, "Adalbert Stifter als 'Conservator,'" 195–196.

reflective hindsight, in his novel *Indian Summer* (1857). But what of his larger body of narrative prose, the works that made him famous: the novellas collected in various volumes of *Studies* and *Many-Colored Stones?*

Like the theme of "nature" or the descriptions of natural landscapes, geological themes are omnipresent in Stifter's fiction and have long attracted the attention of scholars.[84] Indeed, if one understands geology in the broadest sense as "earth sciences," then nearly every fictional work engages with some aspect of the physics or *Naturlehre* that the author learned as a student in Vienna from Andreas Baumgartner, or from other readings on the Tellurian cosmos that constituted the greatest part of Alexander von Humboldt's famous *Kosmos.*[85] Sabine Schneider traces the particular dynamic of "cultural erosion" in the novella *Abdias*, which follows its title character from the ruins of a "Roman city lost to history" in a sandy North African desert to the presumably more stable abode in an Austrian forest – whose stability is belied by the novella's tragic end, however.[86]

---

84 For an overview, see Peter Schnyder, "Geologie und Mineralogie," in *Stifter-Handbuch*, ed. Begemann and Guiriato, 249 – 253. The earliest comprehensive study is by Edgar Banitz, "Das Geologenbild Adalbert Stifters," in *Gestaltung, Umgestaltung. Festschrift zum 75. Geburtstag von Hermann August Korff*, ed. Joachim Müller (Leipzig: Koehler & Amelang, 1957), 206 – 238. See also Lori Wagner, "Schick, Schichten, Geschichte: Geological Theory in Stifter's *Bunte Steine*," *Jahrbuch des Adalbert Stifter-Institutes des Landes Oberösterreich* 2 (1995): 17 – 41, and more recently Georg Braungart, "Der Hauslehrer, Landschaftsmaler und Schriftsteller Adalbert Stifter besucht den Gletscherforscher Friedrich Simony. Hallstatt, im Sommer 1845," in *Bespiegelungskunst. Begegnungen auf den Seitenwegen der Literaturgeschichte*, ed. Georg Braungart et al. (Tübingen: Attempto, 2004), 101 – 118, and Jana Schuster, "Sanftmütige Gärtner, oder: post-katastrophische Kulturation im frühen Anthropozän. Bio- und Geonarrative bei Adalbert Stifter," in *Repräsentationsweisen des Anthropozäns in Kultur und Medien*, ed. Gabriele Dürbeck and Jonas Nesselhauf (Bern: Peter Lang, 2018), 29 – 47.

85 Baumgartner's popular textbook, *Die Naturlehre nach ihrem gegenwärtigen Zustande mit Rücksicht auf mathematische Begründung*, went through several print runs in the 1830s and 1840s. Cf. Moriz Enzinger, *Adalbert Stifters Studienjahre (1818-1830)* (Innsbruck: Österreichische Verlagsanstalt, 1950), 128 – 151; Franz Pichler, "Andreas Baumgartner und sein Werk zur Naturlehre," in *Stifter und Stifterforschung im 21. Jahrhundert: Biographie, Wissenschaft, Poetik*, ed. Alfred Doppler et al. (Tübingen: Max Niemeyer, 2007), 117 – 125. Michael Gamper, *Elektropoetologie: Fiktionen der Elektrizität 1740-1870* (Göttingen: Wallstein, 2009), 253 – 308, offers an insightful reading of how Stifter adapts Baumgartner's teachings on electricity and meteorology. On Humboldt and Stifter, see Franziska Schößler, "Der Weltreisende Alexander von Humboldt in den österreichsichen Bergen: Das naturwissenschaftliche Projekt in Adalbert Stifters 'Nachsommer,'" in *Ordnung – Raum –Ritual: Adalbert Stifters artifizieller Realismus*, ed. Sabina Becker and Katharina Grätz (Heidelberg: Winter, 2007), 261 – 285.

86 Schneider, "Kulturerosionen," 265. Cf. HKG 1.5, 239.

The motifs of ruins and sand, as the end state of ruinous erosion, likewise appear throughout Stifter's work, though they are perhaps most central to his novella *Limestone*.[87] Of the stories collected in *Many-Colored Stones*, *Limestone* is arguably the most "geological." On the whole, however, the story titles symbolically represent the theme of each story: granite for the durability of family tradition; tourmaline for the dark theme of crime; white *Bergmilch* for a story about innocence and heroism; or fool's gold (*Kazensilber*) for a tale of false appearances.[88] Indeed, in one story, *Rock Crystal*, the "crystal" in the story is not even a mineral at all, but rather glacial ice – a category mistake that Stifter nods to in his introduction by recalling how pieces of polished glass often made their way into his childhood mineral collection.

By contrast, limestone or *Kalkstein* is both a symbol and a key plot element in a story that is, moreover, a paradigmatic demonstration of the principle that Stifter claims underlies the poetics of *Many-Colored Stones* as a whole, namely the so-called "sanftes Gesetz" [gentle law]. This law emerges from an understanding that all events in nature are of the same quality or "greatness," given that they are all governed by natural laws. Citing this law, Stifter defends his particular interest in slow, constant and recurrent phenomena, as well as cumulative and hence massive effects, an interest that falls in line with contemporary earth science, be it meteorological research or Charles Lyell's geological actualism:[89]

> Das Wehen der Luft das Rieseln des Wassers das Wachsen der Getreide das Wogen des Meeres das Grünen der Erde das Glänzen des Himmels das Schimmern der Gestirne halte ich für groß: das prächtig einherziehende Gewitter, den Bliz, welcher Häuser spaltet, den Sturm, der die Brandung treibt, den feuerspeienden Berg, das Erdbeben, welches Länder verschüttet, halte ich nicht für größer als obige Erscheinungen, ja ich halte sie für kleiner, weil sie nur Wirkungen viel höherer Geseze sind.

87 Franziska Frei Gerlach, "Die Macht der Körnlein. Stifters Sandformationen zwischen Materialität und Signifikation," in *Die Dinge und die Zeichen: Dimensionen des Realistischen in der Erzählliteratur des 19. Jahrhunderts*, ed. Sabine Schneider and Barbara Hunfeld (Würzburg: Königshausen & Neumann, 2008), 109–122.

88 For a thorough reading of possible symbolism, cf. Wagner, "Schick, Schichten, Geschichte," 33–41.

89 Earlier scholarship attempted to situate Stifter's "gentle law" within philosophical traditions, but considering the author's interest in natural science, a knowledge-historical contextualization seems more germane. Cf. Sepp Domandl, "Die philosophische Tradition von Adalbert Stifters 'Sanftem Gesetz,'" *Vierteljahresschrift des Adalbert Stifter-Instituts des Landes Oberösterreich* 21, no. 3/4 (1972): 79–103.

> The flow of the air, the rippling of the water, the growth of the grain, the waves of the sea, the greening of the earth, the gleaming of the sky, the twinkling of the stars I consider great; the splendidly rising storm, the lightning that splits houses, the tempest that drives the surf, the fire-spewing mountain, the earthquake that buries whole countries, I consider not to be greater than the former phenomena; indeed, I consider them smaller because they are only the effects of much higher laws.[90]

While Stifter is on firm scientific ground when he insists on the universality of "higher laws" of nature, his own, clearly moral preference for "the small" is more a matter of taste. For most critics, however, the more questionable argument is the analogy that constitutes the "gentle law" itself: "So wie es in der äußeren Natur ist, so ist es auch in der inneren, in der des menschlichen Geschlechtes" [As it is in the eternal nature, so it is in the internal nature of the human race].[91] Transferred into the moral realm, the preference for the small or "Einfachheit" [simplicity] soon becomes a plea for "Bezwingung seiner selbst" [mastering oneself] and "Wirksamkeit in seinem Kreise" [tranquil effort].[92] As ethics, the aesthetics of the small resembles a quietist call to 'cultivate your garden'. In a rhetorical parallelism, Stifter also condemns the opposite of this quiet self-control, the "mächtige Bewegungen des Gemüthes" [mighty movements of temperament], and argues that they are "nur Hervorbringungen einzelner und einseitiger Kräfte [...] wie Stürme feuerspeiende Berge Erdbeben" [as much products of individual and one-sided forces as storms, fire-spewing mountains, and earthquakes]. In other words, the analogy not only borrows an understanding of natural law to specify a moral program, but it also naturalizes a specific moral program based on self-control and self-restraint. While Stifter thus defends the "Kräfte, die nach dem Bestehen des Einzelnen zielen" [forces that aim for the survival of the individual] as essential to the "gentle law," he nonetheless subsumes them under the authority of the "Kräfte, die nach dem Bestehen der gesammten Menschheit hinwirken" [forces that work toward the survival of mankind as a whole].[93]As a moral law, the gentle law tolerates individuals, but not individualists; it pretends to abhor the powerful product of "one-sided" forces, but respects the survival of humanity as the law that overpowers all others.

---

90 HKG 2.2, 10. Stifter, "Preface to the Many-Colored Stones," in *German Novellas of Realism: Stifter, Droste-Hülshoff, Gotthelf, Grillparzer, Mörike*, ed. Jeffrey L. Sammons (New York: Continuum, 1989), 1–2.

91 HKG 2.2, 12 [3].

92 Ibid.

93 Ibid.

Limestone demonstrates this moral law through the actions of its protagonist, a country priest who strives to assure the "survival of mankind."[94] The priest's "tranquil effort" is limited to his isolated parish, situated in a rugged Karst landscape. The narrator, a state-employed land surveyor, thus reports on the life of the priest as an addendum, so to speak, to his survey of the region. The geological gaze of the land surveyor thus reveals the central analogy between the priest himself and the limestone-based, karst landscape. Just as the surveyor studies the sediments of rock, so too he observes at their first meeting various layers of the priest's wardrobe:

> Bei den Ärmeln gingen, wie er so saß, manchmal ein ganz klein wenig eine Art Handkrausen hervor, die er immer bemüht war wieder heimlich zurük zu schieben. Vielleicht waren sie in einem Zustande, daß er sich ihrer ein wenig hätte schämen müssen.
>
> Sometimes, as he sat there, a slight trace of his shirt cuffs would become visible, and these he always carefully and surreptitiously pushed back into his sleeves. Perhaps their condition was such that he must have been slightly ashamed of them.[95]

We later learn the secret of this deeper layer of clothing, and the priest is indeed not ashamed of the shabbiness of his cuffs, which would be in keeping with the rest of his appearance. Instead, he is ashamed of the luxury of these white linens that he holds onto from his youth, as a fetish for his unfulfilled love for the daughter of a laundress.

Like the landscape in which he lives, the priest thus has a prehistory hidden in the layers of sediment. The narrator's description of his second encounter with the priest returns to the metaphor:

> Eines Abends, als ich von meinen Arbeiten allein nach Hause ging [...] sah ich meinen armen Pfarrer auf einem Sandhaufen sizen. Er hatte seine großen Schuhe fast in den Sand vergraben, und auf den Schößen seines Rokes lag Sand. Ich erkannte ihn in dem Augenblike.
>
> One evening, when I was walking home alone from my work [...], I saw my poor priest. He was sitting on a mound of sand, with his big shoes almost buried in it, and sand on his coattails. I recognized him immediately.[96]

---

94 Paul Fleming reads "Limestone" as an example of Stifter's poetics of "average" people, which in the statistical sense emerges from the accumulation of large numbers, cf. *Exemplarity and Mediocrity*, 139–162.

95 HKG 2.2, 65; Stifter, "Limestone," in *German Novellas of Realism*, ed. Sammons, 37.

96 HKG 2.2, 68 [39].

Why is it so easy to recognize the priest when he is half-buried in sand? Metaphorically, the priest is the sand. In the eyes of the surveyor, he is as much part of the landscape as any limestone formation he measures and records. For the priest has a prehistory, a layer hidden in sediments of his past, that the narrative will seek to uncover. The priest's actions are also sedimentary: he slowly and meticulously saves money: "Die kleinsten Beiträge, selbst Pfennige, wurden zugelegt, und neue Quellen, die unscheinbarsten, eröffnet, daraus ein kleines Fädlein floß" [he entered the tiniest sums, even pennies, and started the most modest of further economies if they would yield an additional mite].[97] The purpose of this saving, moreover, returns it to the realm of geology. Intended for the construction of a new schoolhouse, so that the local children will no longer have to cross a dangerous river that frequently floods, the savings amount to a preventative measure in the face of environmental danger.

If the priest's cumulative saving is a kind of sedimentation, its aim is thus also to counteract eruptive erosions. The children's path would not be so dangerous were it not for frequent storms and flooding, which provide the dramatic action of Stifter's novella. In short, the priest's moral act of saving is directly opposed to the "splendidly rising storm, the lightning that splits houses" that Stifter abhors in his preface. His self-mastery and thrift does not bring him to personal happiness nor the individual success of a fulfilled marriage, but it does help the entire community to thrive. The actions of the individual may be insufficient, but they inspire a collective effort:

> Als die Sache mit dem Testamente und dessen Unzulänglichkeit bekannt wurde, traten gleich die Wohlhabenden und Reichen in dem Umkreise zusammen, und unterschrieben in Kurzem eine Summe, die hinlänglich schien, alle Absichten des Pfarrers vollziehen zu können.
>
> When the story of his will and the inadequacy of its provisions became generally known, the rich and the prosperous people of that area at once made common cause and before long they had collectively subscribed a sum that seemed quite large enough to fulfill the priest's intentions.[98]

---

97 HKG 2.2, 129–130 [84].

98 HKG 2.2, 131–132 [86]. The detail of the will and testament situates *Limestone* within Stifter's other "testamentary fictions," including *Castle Crazy*, cf. Ulrike Vedder, "Erbschaft und Gabe, Schriften und Plunder. Stifters testamentarische Schreibweise," in *History, Text, Value: Essays on Adalbert Stifter, Londoner Symposium 2003*, ed. Michael Minden, Martin Swales, and Godela Weiss-Sussex (Linz: StifterHaus, 2006), 22–34.

The Madness of Restoration: *Castle Crazy*

Limestone and the other *Many-Colored Stones* represent Stifter's most concerted and conspicuous engagement with geological themes in his work before the geologist novel *Indian Summer.* It is less clear, however, where to look to find the related theme of restoration in his early narratives. As we have seen, the topic interested the author of *Vienna and the Viennese*, and the motifs of ruin and of (agri-)cultural renewal appear frequently in the *Studies. The High Forest* (*Der Hochwald*), for example, employs the ruin as a framing device to explore a historical narrative, while *Brigitta* tells a tale of love and landscape restored by patient cultivation – both threatened, as the text's topical imagery makes clear, by violent magnetic or volcanic forces.

The short novel that most explicitly and most complexly investigates the problematic relationships between geology and history, between material entropy and cultural tradition, bears a title that also may reveal the younger Stifter's frustration with unresolved paradoxes of this discourse: *Die Narrenburg, Castle of Fools* or, in the (witty) contemporary English translation, *Castle Crazy.*[99] The question of what, exactly, is "crazy" about this castle or what makes its inhabitants "fools" goes to the heart of the problems and paradoxes that *Indian Summer* will later attempt to resolve. In a sense, the question is captured in the two possible translations or interpretations of the composite *Narrenburg:* Is it merely a landlocked variant of the trope of the *Narrenschiff* [ship of fools] or is there something about the Castle itself that is or makes one crazy?

On multiple levels, the narrative – and its various answers to this question – unfolds around the difference between present and past, or more pointedly, between life and history. The story begins by expressing this difference as the difference between biological generation and autobiographical genealogy.[100] The principle peculiarity of *Castle Crazy*, namely, can be found in the terms of the "lächerliches Fideicommiß" [ridiculous trust] that regulates the succession of the Duchy to which it belongs.[101] As in other such trusts, ownership of the estate

99 Adalbert Stifter, *Castle Crazy and Maroshely* (London: R. Bentley, 1851). The translation follows the book publication of *Studien* found in HGK 1.4.
100 Cf. Sigrid Weigel, "Zur Dialektik von Geschlecht und Generation um 1800. Stifters 'Narrenburg' als Schauplatz von Umbrüchen im genealogischen Denken," in *Generation. Zur Genealogie des Konzepts – Konzepte von Genealogie.*, ed. Ohad Parnes et al. (Munich: Fink, 2005), 109–124, and more generally, Weigel, *Genea-Logik: Generation, Tradition und Evolution zwischen Kultur- und Naturwissenschaften* (Munich: Fink, 2006).
101 For a legal history of such trusts, see Jörn Eckert, *Der Kampf um die Familienfideikommisse in Deutschland: Studien zum Absterben eines Rechtsinstitutes* (Frankfurt a.M.: Peter Lang, 1992).

"solle sich in gerader Linie immer auf den ältesten Sohn forterben; ist kein Sohn da, auf Töchter, und in Ermangelung dieser auf die älteste Seitenlinie und so fort" [was always to descend in a direct line to the eldest son; should there be no son, to the daughters; and in default of these to the eldest collateral line, and so on].[102] Such arrangements constitute the basic patriarchal form of genealogy: each generation is inscribed in its place within a line or tree. The trust established by the Duke Hans von Scharnast, however, has a further stipulation: each heir must write a detailed autobiography, and read the collected autobiographies of his ancestors, or, failing this, must renounce his or her claim to the estate. In addition to the genea-logical inscription of generation, then, one finds here a bio-graphical supplement. An entire archive is thus created that includes the stories of lives that might otherwise find the sparest inscription in a genealogical tree.

Viewed from today's knowledge of DNA– in which a history of genealogical code is 'inscribed' in each 'heir,' Stifter's construction is striking, and the consequences he imagined for it even more so. The founder believed that the autobiographies would teach each heir how to live a better life, in the model of the exemplary *historia magistra vitae* or "history as a teacher of life."[103] At the very least, he hoped that shame might prevent the commission of sins for fear that, in the near future, they would be committed to paper. Yet in keeping with the satirical ship-of-fools trope, the opposite proves to be the case. The members of the Scharnast family often follow the example of their forebears' crimes; and reading this history of malfeasance appears to eliminate any feelings of shame. That aside, however, the family biographical code *grows* with each generation.

In this respect, and on a purely formal or material level, the clause produces a kind of perpetual growth. The reproduction of the family thus produces an endless archive of its story. This is also the case for the novella's most memorable motif, the castle itself, which is rather "eine Sammlung von Schlössern, eine halbe Stadt von Schlössern" [a collection of castles, half a town of castles], that is, various buildings built by various generations.[104] The castles are physical monuments to the history recorded in the names and texts of the genealogical

As the title suggests, the *Fideicommiß*, which had primarily served to defend the property of noble families from royal and imperial prerogatives, starts to disappear as a legal institution in the mid-nineteenth century, as it is replaced by bourgeois civil law.

102 HKG 1.4, 321 [1].

103 The failure of this clause suggests the "dissolution of the topos" famously discussed by Reinhart Koselleck, *Futures Past: On the Semantics of Historical Time*, trans. Keith Tribe (New York: Columbia UP, 2004), 26–42.

104 HKG 1.4, 328 [12].

and autobiographical archives, "ein Geschichtsdenkmal, wie ein viele Jahrhunderte altes, vielleicht närrisches, vielleicht geniales Geschlecht wunderbar auf diesem wunderbaren Berge gehauset und gegraben und gebaut hatte" [an historical monument, as if built by the wonderful living and digging, on a wonderful mountain, of a centuries-old race that was perhaps foolish, perhaps genius].[105] While they both represent the family's past, the autobiographical stories and the architectural history differ in one key respect – key, that is, to Stifter's developing understanding of history primarily as a history of material things. First, when the text finally cites, in its last chapter, an excerpt from the autobiographical archive, it contains a scathing self-reflective criticism of the life-writing practice of the family:

> der Dämon der Thaten steht jederzeit in einer neuen Gestalt vor uns, und wir erkennen ihn nicht, daß er einer sei, der auch schon euch erschienen war – und eure Schriften sind mir unnütz. Jedes Leben ist ein neues, und was der Jüngling fühlt und thut, ist ihm zum ersten Male auf der Welt: ein entzückend Wunderwerk, das nie war, und nie mehr sein wird – aber wenn er vorüber ist, legen es die Söhne zu dem anderen Trödel der Jahrtausende, und es ist eben nichts als Trödel; denn jeder wirkt sich das Wunder seines Lebens aufs Neue.
>
> The demon which prompts our actions comes to each of us in a new form, and we cannot recognize him as the same which tempted you, and therefore your writings are useless to me. Every life is a new life, and whatever the youth feels and does, is to him as though it had never been felt or done before. And when all is over, each adds his share to the rubbish (*Trödel*) of past [millennia] which fills this chamber – and nought is it but rubbish, for every one works the mystery (*Wunder*) of life afresh.[106]

Life is here defined strictly as present life; each individual life is the "Schlußstein des millionenjährig bisher Geschehenen und der Mittelpunkt des All" [key-stone to the millions of years that have been, and the central point of the universe], until, of course, the next life succeeds it.[107] In this way, the writing of life, which, like the *historia magistra vitae*, hopes to transfer life lessons from one generation to another, is nothing but "rubbish" (*Trödel*).

Stifter's text associates this understanding of "life" with a place, the Green Fichtau, the local inn where the story's protagonist Henry stays and meets his future wife, the innkeeper's daughter Anna. As the color "green" suggests, the location is idyllic, and life there follows the cyclical rhythms of nature, whose repetition creates the impression of an eternal present. By contrast, the castles are called the "Red Rock," because of the red marble used in its construction.

105 My translation of this sentence from the first, "journal" edition: HKG 1.1, 348.
106 HKG 1.4, 410 [124].
107 HKG 1.4, 410 [125].

Just as the marble points to a deep geological past, and the irretrievable linearity of earth history, so too do the castle ruins suggest the unstoppable march of historical time. Stifter expresses this distinction most pointedly in a passage that characterizes the Fichtau valley as "grün und weich und ruhig [...] eine holde Gegenwart" [green and peaceful as ever [...] sweet happy Present], and contrasts it sharply with the castle ruins, "herumliegend um die unklare Vergangenheit, auf der sie standen" [encircling the dark and troubled Past]. The contrasting motifs are evident throughout Stifter's long descriptions of both the valley's vivid natural beauty, and the castle's dead, rocky, ruinous state.[108]

In addition to such topical distinctions, Stifter's text elucidates the difference between the present, at the inn, and the past of the castle as a difference in media. In the country inn, all communication is oral, and as a result, all history is uncertain and to some extent, unimportant. On the positive side, this means that the green present is also eternal, the space and medium in which the "mystery of life" always works "afresh." On the negative side, this means that none of the accounts are legally or historically binding, including the account of a fraternal rivalry between Julius and Julianus. This dispute, which causes a split in the family tree, constitutes the narrative's driving secret – revealed only when Henry recognizes and then confirms his identity as a descendent of Julius, the disinherited younger brother of Julianus.

As the similarity of the names Julius/Julianus, or more obviously, other ancestors such as Sixtus I and II or Henry I and II, make clear, oral history also has difficulty distinguishing between figures who appear as different only in the writing of the archive. Henry only becomes "Heinrich II" once archival research confirms his lineage. As the text makes clear in a second stage of revealing Henry's past, in the castle's portrait gallery, an image-based or visual archive of family history has the same problem as oral history. Physically, Henry resembles the Schnarast family, and even presents the spitting image of his relative Sixtus II,

108 HKG 1.4, 368–369 [68]. An earlier passage evokes a similar contrast between present and past: "Allein der Zweck der vorliegenden Blätter führt uns aus dieser harmlosen Gegenwart, die wir mit Vorliebe beschrieben haben, einer dunklen schwermütigen Vergangenheit entgegen, die uns hie und da von einer zerrissenen Sage, oder einem stummen Mauerstücke erzählet wird, denen wir es wieder nur eben so dunkel und mangelhaft nacherzählen können" HKG 1.4, 361 [The object of our pages carries us away from this happy and guileless present, which otherwise we should far prefer picturing, to a dark and gloomy past, which, gathered mostly from the fragmentary traditions of old and mute walls, it is our task to describe as best we can, and best will be but obscure and defective, 58]. The narrator's self-reflection suggests an equivalence between the writer and the restorator with respect to the fragments of the past.

whom the old castellan still knew in his youth. Yet such physical resemblance has no legal authority.

The portraits do have historical authority, however: not in the faces they portray, but in the historical costumes and art-historical styles they present.[109] In this regard, the images constitute one element in the historical archive of the "Red Rock," which also includes the buildings, and indeed the "rock" itself. It is no coincidence that Henry, the contemporary heir who promises to restore the family line, was "in die Fichtau gegangen, um Blumen und Steine zu sammeln, und ha[t] das todte Geschlecht [s]einer Väter gefunden" [visiting the Fichtau merely to pick up flowers and stones, [...] discovered the ancient heritage of [his] fathers].[110] It is his "knowledge of stones" that allows him to find the castle in the first place, discover its meaning, and ultimately, to restore *it* as much as the family. When Henry, whom we have already encountered with "ein langstieliger Hammer, und eine abenteuerliche Hacke [...] als trüge er beide Säcke voll Eisen oder Gestein" [long handled hammer and an axe [and pockets] full of iron and stone], first discovers the castle, his curiosity is most struck by his inability to find an entrance.[111] The image of the sealed, gateless castle suggests a hermeneutic secret, but this lack of gate also makes it indistinguishable from a natural rock formation. The "Red Rock" is, first and foremost, a rock.

As a naturalist, Henry is uniquely qualified to understand this bizarre "complex" of nature and history.[112] From the grey, stony face of the old castellan Ruprecht, to the cellular beehives that have taken over the "cells" of a nunnery, to the sand that has overtaken and ruined an obelisk monument, the text is replete with images that suggest an equivalence between nature and culture.

Indeed, the entire plot of Henry's discovery of his heritage and happy marriage to Anna revolves around a botanical metaphor. The text compares the innkeeper's daughter to a rare Alpine flower, whose happiness depends on the right environmental conditions. Propelled by Alexander von Humboldt's *Geography of Plants*, this ecological insight finds confirmation in Stifter's text through a neg-

109 Stifter "dates" the paintings in the portrait gallery by naming famous painters such as Rubens, Van Dyck, Murillo, and Dürer, but also by describing changes in fashion, HKG 1.4, 382–386 [87].

110 HKG 1.4, 400 [110]. The translator effaces the fact that the "heritage" is as "dead" as the stones.

111 HKG 1.4, 324 [5].

112 In the journal edition, Stifter employs his usual parataxis to underscore the inseparability of culture and nature here: "Complex dieser Bauwerke und Gärten und Wälder," HKG 1.1, 350. The book edition reads here, "diese ganze weitläufige Mischung von Bauten, Gärten und Wälder," HKG 1.4, 367 [these varieties of buildings, gardens, and woods, 65–66].

ative case, which appears in one of the novella's internal narratives. If Anna thrives because she is well adapted to the Alpine climate at the Green Fichtau, Chelion, the wife of an earlier Scharnast, who is described as an exotic "flower" from India, suffers and dies tragically as a consequence. In this way, Stifter's novella explains migration and the ensuing cultural conflict as a matter of botanical transplantation.[113] The figure of Chelion, moreover, has a historical basis in the thirteen-year-old Ethiopian girl Machbubah whom Fürst von Pückler-Muskau married and introduced to Vienna society in 1837.[114] By naturalizing this intercultural encounter, widely sensationalized in the Viennese press, Stifter's text provides the basis for a racial interpretation of its "failure" – and thereby echoes the racism and sexism surrounding the public scandal of the marriage. By extending the commonplace metaphor of woman as flower, Stifter's text exposes its more troubling, dehumanizing implications: while the male European naturalist can travel the world, the woman he finds and "picks" to take home will suffer the consequences.

The botanical metaphor, moreover, also points to Pückler-Muskau's fame as a horticulturalist, which opens up the possibility of reading the entire castle complex as a Romantic garden in the style of Pückler-Muskau. Stifter's text refers explicitly to "seine romantischen Vorfahrer" [his romantic ancestors].[115] This attribution places the novella's protagonist in the position of the post-romantic, looking back at a period known for its fascination with ruins. It also contains a hidden critique of Pückler-Muskau's Romantic brand of historical culture. His fictional counterpart Jodok, for example, constructs a Grecian temple. Such a building is not only out of place in the Alpine Fichtau, but also falsely imports an exoticism that only vaguely suggests Chelion's Indian home. Stifter adds additional inaccuracies to characterize this inauthentic exoticism: Chelion is in fact Hebraic, a Biblical name, and, in the first, journal edition of *Die Narrenburg*, the "Parthenon" building is described as looking like a mosque as well as an Indian garden.[116] The castle and garden complex thus includes both gen-

113 Cf. Steeg, *Wissenskunst*, 181–183.

114 See Ulrich Dittmann's commentary in HKG 1.9, 208.

115 HKG 1.1, 351.

116 "fast wie eine Moschee aussehend, aber statt der Wände waren nur schlanke Säulen mit breiten Zwischenräumen zur Umsicht" [almost looking like a mosque, but instead of walls there were only thin columns with wide spaces in between for better visibility], HKA 1.1, 352. Stifter's description hints cleverly at the double displacement that occurs in historicist building: a model is not only taken from a foreign time or place, but is also changed to serve a new function. On the historicist garden, see Wolfgang Schepers, "Zu den Anfängen des Stilpluralismus im Landschaftsgarten und dessen theoretischer Begründung in Deutschland," in: *Geschichte ist allein zeitgemäß*, ed. Michael Brix and Monika Steinhauser (Lahn-Giessen: Anabas, 1978), 73–93.

uinely historical buildings and more recent, historicist recreations, such as Jodok's Hellenistic "Parthenon," but also Christoph's palace: "Das Gebäude, obwohl neu, war doch nicht in neuem Style gebaut, aber auch nicht im alten" [The building, although it was new, was not built in the new style, but also not in the old one].[117]

The failure of Jodok and Chelion's marriage, and the burning of the inauthentic temple he constructed, critiques a certain type of historicism – and the happy union of Henry and Anna suggests the promise of another. Hans-Georg von Arburg identifies Henry's restoration project as a "third way" between historicism, inaugurated by Heinrich Hübsch's question, "In welchem Style sollen wir bauen?" [In which style should we build? 1828], and the more modern architecture of Gottfried Semper. The latter published his main theoretical work *Style in the Technical and Tectonic Arts* in 1860 – 1863, but Arburg sees the resonance of Semper's theory (and earlier practice) in the "aesthetics of autonomy" that Stifter inherits from the age of Goethe and expresses in the "hermetic outer wall."[118] The wall, however, is not hermetic in one important respect: it is open towards nature – because it is nature:

> das ganze durch eine klafterdicke, hohe Granitmauer umfangen, an der, nach der Bauart zu schließen, viele Geschlechter gearbeitet hatten, und die er thörichter Weise bei seiner Entdeckung des Schloßes für die äußere Mauer desselben gehalten, weil er unten und dicht an ihr herumgeklettert war.
>
> the whole surrounded by a six-foot thick (*klafterdick*), high granite wall, on which, to judge from the way it was built (*Bauart*), many generations (*Geschlechter*) had worked, and which he foolishly thought at first was the outer wall of the castle, as he was climbing down below and right up against it.[119]

At issue in this passage is once again the ambiguity of Henry's "foolishness" (here: *thöricht*): in what way is the outer wall *not* the outer wall? In Arburg's reading, the singular outer wall precariously holds a multiplicity of castles within, and Henry's mistake was to think that there was only one castle. Yet if "six-foot thick" (*klafter* is an archaic unit of length equivalent to a man's arm span) is not a hyperbole, then the granite "wall" may truly be something that generations have "worked on," but which already existed as a natural formation. Indeed, just

---

117 HKG 1.1, 353.

118 Hans-Georg von Arburg, "Neues von der *Narrenburg?* Stifters Architeckturen zwischen Historismus und Neuem Bauen" in: *Figuren der Übertragung*, ed. Gamper and Wagner, 133.

119 HKG 1.1, 350. The passage in the book edition attempts to remove the ambiguity by calling it an "iron wall" or "Eisenmauer," HKG 1.4, 367. Significantly, the book version also eliminates the passages about Henry's restoration plans.

such a natural wall surrounds the ruins of the Castle Scharnstein, south of Stifter's boarding school in Kremsmünster, which served as inspiration for the castle in the story. At this "archaic border between nature and culture," only a geologist can discern where the *Bauart* of the cliff ends – the term is a frequent metaphor for natural "constructions" – and the style of the castle begins.[120]

Henry, the expert of plants and stones, thus develops plans to "restore" the castle in the most modern way, that is, by respecting its authentic, material history. When Henry surveys from a mountaintop the castle of his ancestors, he discovers a "Geschichtsdenkmal" [historical monument] that reveals centuries of historical change.[121] His heart races at the thought that he will inherit the "Complex" of buildings, gardens, and forests, and his head begins to spin with ideas for renovation and restoration:

> [...] ein unabsehliches Feld von Wirken und Schaffen that sich ihm auf, und von Genießen und Schwärmen für seine arbeitende närrische Phantasie, der ein solches Dichtungschaos, wie dieser Berg, gerade willkommen war. Er mußte von der Platte weggehen; denn ein Wirrsaal und ein Getümmel von Plänen und Gedanken kam schon in sein Haupt, die sich schoben und drehten und noch keine Gestalt hatten.
>
> an incalculable field of working and creating appeared before him, and of enjoyment and enthusiasm for his working, foolish fantasy, to which such a chaos of poetry like this mountain was so welcome. He needed to leave this place; for a confusion and tumult of plans and thoughts was already entering his mind, and they pushed each other and spun and refused to take any specific form.[122]

Just like the archive of family autobiographies that lie stored within, the castle ruins themselves are a "Dichtungschaos" [chaos of poetry] that inspires Henry's "arbeitende närrische Phantasie" [industrious foolish fantasy]. Like his forebears, then, Henry Scharnast is thus a fool (*Narr*), but his foolish fantasy differs from theirs in an important respect. While it does continue the history of change that produced the bizarre complex, it also surveys this history and sees its mission in the conservation and continuation of the traditions they left behind. The task is enormous, of course, "an incalculable field of working and creating." The young man immediately wishes that he had three hundred years to live and complete it. At the same time, he already feels that he is part of the family: "manches bereits Bestehende sah ihn so bekannt an, als hätte er es selbst vor hundert Jah-

120 Arburg, "Neues von der Narrenburg," 115, including a woodcut of Scharnstein.
121 HKG 1.1, 348.
122 HKG 1.1, 350.

ren begonnen" [much of what was already here looked so familiar to him that it seemed as if he had begun work on it himself a hundred years ago].[123]

While *Castle Crazy* does not provide a theory of restoration, it does suggest that Stifter imagines restoration as emerging within the context of geology. It is no coincidence that the castle (*Burg*) Henry inherits and wishes to restore is also a mountain (*Berg*). Henry himself is a naturalist and not an art historian in the classic sense. Just as Henry discovers the castle while on a geological survey, his conservation efforts are as intensely devoted to the garden and forest as they are to the buildings. Geohistoricism does not distinguish among the family, its castle, or the mountain on which that castle rests: "Der ganze Berg erscheint mir, wie eine Tradition, davon ich selbst ein Glied bin" [the whole mountain appears to me as a tradition, of which I myself am a member].[124] The protagonist of the novel *Indian Summer*, another geologist named Henry, will understand this sentiment. As a geologist, he will imagine that every mountain is a tradition of which he is part. The fact that the latter Henry is also the first-person narrator of his own story suggests, moreover, that the writer Stifter increasingly identifies with the feelings of the geological *Conservator*. By contrast, *Castle Crazy* ends with the narrator's rather "crazy" promise to publish more stories from the history of the Scharnast family. Stifter pursued this editorial fiction to the end of his writing career as a project with the title *Die Mappe meines Urgroßvaters* [My Great-Grandfather's Notebook], though he did not produce the serial installments as the narrator of the *Narrenburg* promises. Instead, he wrote and rewrote four versions of the *Notebook*. In the third and fourth versions of this text, the fictional "notebook" in question bears the title: *Calcaria Augustiniana Tom. II.* Stifter's commentators have read "calcaria" as the plural of "calcar" [spur], derived from "calx" [heel].[125] Yet given the ubiquity of stones in Stifter's work, it is hard not to read here the homonym "calx" [lime] and its adjective "calcarius:" of or pertaining to limestone. Read as "the Augustinian things of limestone," the moniker imagines anthology not as a "collection of flowers," nor even as a collection of many-colored stones, but as a sediment. Instead of a series, then, Stifter ultimately produced – not least through his rewriting of the same text – in the *Notebook* a literary sediment, and in *Indian Summer*, he wrote the ultimate literary guide for reading and restoring the sediments of history.

---

123 HKG 1.1, 351.

124 HKG 1.1, 352. The panoramic view of the castle complex, in turn, recalls the geological panoramas in Stifter's other works such as *Vienna and the Viennese* and *Indian Summer*, cf. Schneider, "Kulturerosionen," 253–259.

125 Silvia Bengesser and Herwig Gottwald, "Stellenkommentar," HKG 6.4, 254.

Chapter 2 The Summer of Restoration

Pure Posteriority

Stifter's novel *Der Nachsommer* [*Indian Summer*, 1857] bears the inscription of temporality in its title. The word refers both to a period of time and a structure that relates one period to another: it designates the "Indian summer" itself, a warm autumn that is like the summer, and the *Nach-sommer*, what comes *after* the summer. An Indian summer is a season out of order: warm weather returns when it is least expected.[1] When the word appears in the novel, its temporality proves to be even more complex: "So leben wir in Glück und Stettigkeit gleichsam einen Nachsommer ohne vorhergegangenen Sommer" [Thus, we are living in happiness and with a sense of constancy as if in an Indian Summer without the preceding summer].[2] This is how the Freiherr von Risach, the elder of the novel's two main characters, describes his current life. It is a stable present, but it is a present marked by the passage of time, by the posteriority of the prefix *nach*.[3] The present is suffused with a virtual past, a past which itself never existed. The Indian summer without a preceding summer is a construction of pure posteriority: *posterior sine priori*.[4]

As discussed in the introduction, this aesthetics of posteriority or belatedness is related not only to the Freiherr von Risach's "restorative utopia," but also to the young Heinrich Drendorf's geological studies. Even though the extensive research literature on Stifter's novel has addressed both of these themes, it has until now treated them separately, or even seen them as antagonistic forces. In his recent summary of the current scientific-historical research on *Indian Summer*, for example, Michael Gamper sees Drendorf's geology as a modern cor-

---

1 Such weather is mentioned at least twice in the novel, where it becomes one of its "predictable" themes: "Das vorausgesagte schöne Wetter war eingetroffen" [the predicted beautiful weather had arrived], HKG 4.3, 87. Cf. the warm fall weather at HKG 4.2, 135.

2 HKG 4.3, 224 [445].

3 See introduction.

4 As Dieter Borchmeyer points out, if one compares the novel to historical realities such as the post-1848 land reforms, one sees that the plot includes present elements in the past that it purportedly represents: "Die Zeit der Handlung sind, so könnte man paradox formulieren, die in den Vormärz verlegten fünfziger Jahre" ("Stifters Nachsommer – eine restaurative Utopie?" 63). In other words, the novel is emphatically post-1848, even as it reconstructs an alternative reality that takes place in the 1820s and '30s.

https://doi.org/10.1515/9783110685534-004

rective to "classical-restorative" aesthetics represented by Risach.[5] Within the novel itself, however, the relationship between these two figures is much more complex than mere antagonism. From the start, Risach appears interested in and remarkably well informed about his new friend's geological research. For his part, Drendorf, despite a moment of hesitation, appears eager to learn about classical aesthetics and restoration from his older mentor. To judge from the telos of the novel, moreover, the young man's forward-looking research is clearly subsumed into the structure of tradition. As uncertain as the outcome of this research is, the "ground" of family life is stable:

> Ob ich es nun in der Wissenschaft, der ich nie abtrünnig werden wollte, ob mir Gott die Gnade geben wird, unter den Großen derselben zu sein, das weiß ich nicht; aber eines ist gewiß, das reine Familienleben, wie es Risach verlangt, ist gegründet, es wird, wie unsre Neigung und unsre Herzen verbürgen, in ungeminderter Fülle dauern, ich werde meine Habe verwalten, werde sonst noch nüzen, und jedes selbst das wissenschaftliche Bestreben hat nun Einfachheit Halt und Bedeutung.
>
> Whether I shall be able to accomplish anything in science – which I hope never to neglect – whether God will grant me the grace to be numbered among the greats in the field, I don't know; but one thing is certain: the pure family life that Risach was referring to has been founded, it will continue undiminished as our love and our hearts attest; I shall administer my property, shall be useful in other ways, and everything, including my scientific endevours, has now gained significantly in clarity, solidity, and importance.[6]

This conclusion takes care not to reject the geological science of the novel – "which I hope never to neglect" – but it demonstrates an obvious will to contain it, to place it a context where it will receive "importance" or "meaning" (*Bedeutung*). While Drendorf calls this overarching framework "pure family life," it is evident to any reader of the novel that his friend Risach plays a major role in helping the young man find this happiness.[7] One goal of this chapter is thus to explore the relationship between these two figures in order to understand, on a thematic level, how Stifter connects the discourses of geology and restoration they respectively represent.

---

5 Michael Gamper, "Adalbert Stifter: 'Der Nachsommer' (1857)," in *Literatur und Wissen: ein interdisziplinäres Handbuch*, ed. Roland Borgards et al. (Stuttgart: Metzler, 2013), 373. By contrast, Christian Begemann reads the novel primarily from Risach's restorative perspective, *Die Welt der Zeichen*, 321–350.

6 HKG 4.3, 282 [478–479].

7 On the ideal of "purity" and its conflict with the "residues" of geology and history, see Barbara Thums, "Adalbert Stifters 'Der Nachsommer': Reste-lose Poetik des Reinen?," in *Was übrig bleibt. Von Resten, Residuen und Relikten*, ed. Barbara Thums and Annette Werberger (Berlin: trafo, 2009), 79–97.

Indian Summer, however, is not only a story about the reconstructions; it is itself a reconstruction. As Stifter writes to his editor Heckenast in the year of the novel's publication, the story takes place "nicht in unsern Tagen, sondern um dreißig und mehr Jahre zurück" [not in our time, but around thirty years or more ago].[8] (Given the paucity of historical references in the book, it is easy to understand how this hint that it may be a historical novel has largely been ignored. At the same time, it is hardly surprising when one considers that Stifter had long harbored the wish to write historical novels, an intention that also resulted in *Witiko*.[9]) As a historical novel, moreover, *Indian Summer* not only demonstrates a knowledge of the historicity of the present – a criterion for its modernity – but also a knowledge of the historicity of its own historicism. The key historical index in the novel is not only Drendorf's geological research – which more or less represents the state of the art in 1857 – but also Risach's speculation that the work of restoration will one day be carried out by a State Commission. Stifter thus fictionalizes and historicizes his own conservation work. The novel is both a re-imagining of the author's interests in geology and art history, and a commemoration of how these discourses participate in a more general re-imagining of history that includes the literary form of the historical novel.

Following the exploration of the themes of geology and restoration in Stifter's earlier work, I thus read *Indian Summer* as a self-reflective historical novel about the varied discursive underpinnings of its own aesthetic historicism. This reading involves a number of steps. First, an analysis of the novel's narrative structure and the relationship between its two main figures will show how fundamental loss and reconstruction are for the novel as a whole. The next steps consist in showing how the theme of geology is developed, especially as regards its impact on Drendorf's cultural and aesthetic education. The theme of discovery or uncovering not only unites the novel's interpersonal and educational plots, but it also characterizes Stifter's geological aesthetics. The author's famous descriptions of nature describe nature as it is, was and will be. For this reason, Stifter pays particular attention to phenomena that are not only visually spectacular, but which reveal hidden forces at work, in the present or past. In this way, Stifter's text finds multiple figurations for the latent history of the earth that is visible, to the trained eye, in geological formations. Finally, an evaluation of the discussions of restoration and art history in the novel will show the extent to which the novel's historicism is itself geologically motivated. In restoration practices, the virtual reconstruction of earth history becomes a concrete

8 Letter to Heckenast, March 22, 1857, PRA 19, 14.
9 See Chapter 4.

project of reconstructing human history. In the process, human history itself becomes geological, a history that is best understood from the material it has left behind.

Bildungsroman à deux: The Narrative Structure of Restoration

Despite the book's three volumes and several hundred pages, the plot of *Indian Summer* can easily be summarized. In the later years of his self-directed general education, Heinrich Drendorf, a young man from a bourgeois household in Vienna, begins to take longer and longer trips into the mountains of the Austrian Alps. During these wanderings, he develops an interest in the science of geology. One late summer, passing through the hill country on his return from the mountains, he comes upon a remarkable house on a hill – it appears to be "[das] Landhaus[] eines Städters [...] noch hell beleuchtet, und sah mit einladendem schimmerndem Weiß in das Grau und Blau der Landschaft hinaus" [a country estate of someone from the city [...] bathed in sunshine [it] gazed out into the blue and gray of the surrounding countryside with a beckoning glittering white] – just as a thunderstorm is threatening.[10] Seeking refuge there, he makes the acquaintance of its owner, who will formally introduce himself years later as the Freiherr von Risach, after the two have become good friends. The rest of the plot follows Drendorf on his journeys between city and mountains, now with the additional stopover at the house he comes to call the "Rose House" (*Rosenhaus*) for its magnificent rose garden. During these visits, he meets Risach's friends, including a young woman named Natalie, whom he later marries.

What fills the pages of this novel is not twists and turns of plot, but rather extensive description and, as Drendorf's social interactions intensify, long conversations about art and science. Indeed, abstract as the chapter headings are, they give a good sense of Drendorf's smooth life story: "Die Häuslichkeit [...] Der Wanderer [...] Die Begegnung [...] Die Erweiterung [...] Der Einblick [...] Der Bund [...] Die Entfaltung [...] Der Abschluss" [Domesticity [...] The Wanderer [...] The Meeting [...] Broadening the Horizons [...] The Insight [...] The Bond [...] The Unfolding [...] The Conclusion].[11] Each of these titles refers to stages of an initiation and socialization: departure from the childhood home, meeting

10 HKG 4.1, 46 [30 – 31]; On the relative lack of plot in Stifter's novel, and its relationship to the non-event of the storm, see Marianne Schuller, "Das Gewitter findet nicht statt oder die Abdankung der Kunst," *Poetica* 10, no. 1 (1978): 26 – 52.

11 Table of Contents in HKG 4.1, 7; HKG 4.2, 7; and HKG 4.3, 7 [7].

new people, learning new things ("expanding," "developing," and "examining" oneself), and finally, binding oneself to a new family through marriage. If the first sentence of the novel – "Mein Vater war Kaufmann" [my father was a merchant] – is a clear signal of its adherence to the paradigm established by Goethe's *Wilhelm Meisters Lehrjahre* (*Wilhelm Meister's Apprenticeship*, 1796), then these chapter titles suggest a precise schematic analysis of the classical nineteenth-century concept of *Bildung:* an education from and by the book.[12]

Two chapters, however, stand out because they cannot be ascribed to Drendorf, the presumed subject of the *Bildungsroman:* "The Communication" (*Mitteilung*) and "The Look into the Past" (*Rückblick*). The subject of these chapters, rather, is Risach, who appears to usurp Drendorf's role in these two crucial functions of self-expression and retrospection. What Risach communicates in these chapters – which take the form of an extended first-person narration – is precisely what is missing from Drendorf's unusually smooth educational path.[13] At the heart of Risach's autobiographical account lies an irretrievable loss. In his youth, while working as a private teacher for a wealthy family, he falls in love with their daughter Mathilde. Though she shares his feelings, her parents forbid marriage because of her young age. By supporting the parents' injunction, Risach loses Mathilde's respect, which makes any future relationship between the two impossible. Risach's life is thus marked by loss and resignation. Furthermore, as he explains at length in his story to Drendorf, his career in civil service, while not a failure, was not his true calling.

Structurally, these experiences qualify *Risach* to be the hero of a *Bildungsroman*, the genre that marks the entrance into adulthood with a dramatic loss and at least one false step on the path towards a career – or, in the case of Wilhelm Meister, "Irrthümer auf Irrthümer, Verirrungen auf Verirrungen" [mistakes upon mistakes, errors upon errors].[14] It is thus Risach, and not Drendorf, who, like Wil-

---

12 HKG 4.1, 9 [10]. Stifter's novel thus appears in many critical treatments of the genre. Michael Minden, for example, sees in Stifter the expression of the two main (familial) themes of the Bildungsroman "in their clearest forms," *The German Bildungsroman: Incest and Inheritance* (Cambridge: Cambridge UP, 1997), 4.

13 Cornelia Zumbusch reads the contrast between the continuity of Drendorf's story and the discontinuity of Risach's – both in terms of content and in its form as a narrative analepsis, "Nachgetragene Ursprünge: Vorgeschichten im Roman (Wieland, Goethe, Stifter)," *Poetica* 43 (2011): 289–296, and interprets this narrative structure as an expression of nineteenth-century experience of historical change as accelerating, making the past increasingly "obsolete," 271. By integrating geological time into its narrative ductus, Stifter's novel does decelerate time, but it does not bring it to a halt or create the sense of "timelessness," 272, and its restorative poetics wishes, if anything, to bring the past back into the present.

14 Cf. Goethe, *Wilhelm Meisters Lehrjahre*, in: *Sämtliche Werke*, vol. 9, 823.

helm Meister, can only begin his career if he renounces his first love. The most dramatic scenes of the novel take place in Risach's "look into the past," not during Drendorf's wanderings. In *Indian Summer*, the supposed subject of education, Drendorf, makes no mistakes and suffers no loss.

Yet Risach's story does more than interrupt and compete with the successful educational career of his younger friend: it provides the very foundations of the world into which Drendorf enters as a young man. Symbolically, Risach's life functions as a substitute for the past Drendorf never has himself. The younger man learns resignation and self-knowledge by proxy. Listening to Risach's story, Drendorf is thus in the same position as the reader of the novel. The story of learning from love and loss is also prefigured in books that existed before *Indian Summer*. Stifter expresses this intertextual structure by portraying the young Risach and Mathilde above all as readers themselves. Risach first meets Mathilde as a reader: "und seitwärts an einem kleinen Tischchen saß ein Mädchen und hatte ein Buch vor sich. Es schien, es habe vorgelesen" [and to the side a girl was sitting at a small table with a book opened in front of her. It seemed to me she had been reading aloud].[15] And when he falls in love, he turns to his favorite poet: "Ich nahm ein Buch eines Lieblingsdichters, las sehr lange, und feurige Thränen wie heiße Tropfen kamen öfter in meine Augen" [I selected a book by one of my favorite authors and read it for a very long time, with fiery tears that burned like hot drops often coming to my eyes].[16] Such outbursts of emotion are rare for Drendorf, who appears to react to Risach's story pragmatically. The young man's life mimics that of his older friend as its happy variation: he marries his first love, Mathilde's daughter (!), and pursues his passion for science with neither impediment nor doubt.

Risach's influence on Drendorf's life is more concrete, however. Wonder and admiration for Risach's estate and house occupy a good part of Drendorf's narration, things which the young man will indeed inherit through his marriage to Natalie. As we learn from Risach's account, however, the Rose House was, in fact, inspired by the garden house where he fell in love with Mathilde. Not only is this house a memorial to his past love, but its construction also involved extensive renovations of the estate, including the transplantation of the garden from its former location at the bottom of a hill to the hilltop, where Drendorf first sees it, like a beacon attracting him to his fate, "mit einladendem schimmerndem Weiß" [with a beckoning glittering white].[17] If Risach's house is a recon-

15 HKG 4.3, 166–167 [413].
16 HKG 4.3, 191 [427].
17 HKG 4.1, 46 [31].

structed paradise that bears traces of his fall, it is also a historical museum, a museum *of* memorials, full of restored furniture and works of art: things of the past reintegrated into the present.[18] In Risach's museum-like house – visitors to the Rose House wear felt slippers to protect its marble floors – the past becomes present in a way that appears to erase the loss that characterizes it as past.

The Indian summer or "post-summer" (*Nachsommer*) of Risach's life appears to Drendorf as second nature. Unaware of the loss that is its motivation and origin, Drendorf only sees its perfection, indeed, its simplicity and naturalness. The young man, however, is a geologist.[19] In his study of the earth and its history, he comes to the startling discovery that nature itself has a past that hardly resembles its present. His studies reveal to him all sorts of massive changes on the earth's surface: valleys transform into lakes and back into valleys; mountains rise and are worn away; forests grow and die. His conclusions are epoch-making:[20] First, the earth's history is truly historical, i.e., linear and contingent; and second, the position of humanity in this history – his own position – is highly marginal, perhaps even accidental. As a geologist, Drendorf discovers that nature itself is neither an eternal present nor a cyclical "return of the same."[21] The young man's discovery that Risach's world of "second nature," the artificial paradise of the Rose House, also has a past, marked by loss, change, and discontinuity, represents, on a smaller scale, the more radical discovery of the instability of the earth as a whole.

Like Risach's loss, the practice of geology motivates countless reconstructions and restorations. For the geologist, every structure in nature is a relic of an-

---

18 Cf. Heinz and Hannelore Schlaffer, *Studien zum ästhetischen Historismus*, 114. On the museum-like quality of the house, see also Grätz, *Musealer Historismus*, esp. 104.

19 Erhard Banitz, "Das Geologenbild Adalbert Stifters," 206–238, is probably the first study devoted to this topic. Banitz not only identifies Drendorf with Stifter's friend, the geologist Friedrich Simony, 234–236, but also correctly sees that Drendorf is by no means an "amateur," nor does he waver in his calling: "Vollendung als Künstler war nicht Heinrichs Bestimmung, wohl aber sollte die Kunst, soviel er erringt in ihr, den Wissenschaftler vollenden helfen," 233.

20 Few scholars have emphasized that the geological knowledge in *Indian Summer* is historically contemporary with the novel: one is Peter Schnyder, "Schrift – Bild – Sammlung – Karte. Medien Geologischen Wissens in Stifters *Nachsommer*," in *Figuren der Übertragung*, ed. Gamper and Wagner, 235–248, and "Die Dynamisierung des Statischen," 540–555. Also Schneider, "Kulturerosionen," 249–269; and Michler, "Vulkanische Idyllen," 472–495.

21 The notion of nature as an eternal cycle is a powerful topos: H. and H. Schlaffer can thus write, "Wiederkehr des Gleichen aber ist nicht das Wesen der Geschichte, sondern der Natur" (the return of the same is not the essence of history, but of nature) *Studien zum ästhetischen Historismus*, 114, thereby mistaking the central point of Drendorf's geological research.

other time. The very ground becomes a ruin. The geologist's efforts to reconstruct or restore past landscapes may not always be physical; more often, the geologist's work entails image-making and description to recreate past worlds.[22] Drendorf's career as a draughtsman and painter testifies to the similarity between geology and other forms of reconstructive practices in the novel, especially the practice of art restoration. Beginning with sketches of natural objects – plants, animals, and minerals – Drendorf eventually moves on to draw antique furniture, architecture, and other art objects that Risach has collected. In both cases, the drawings constitute a collection or archive of graphic forms that supplements the physical collections, which are likewise made in both domains, but remain, for practical reasons, less comprehensive. These graphic archives thus serve as important source material for imagining the past. In both practices, reconstruction takes on the form of cartography. Every object is marked with its point of origin and trajectory, such that the past becomes visible as a history of movement and displacement.

Like the figures that represent them, the discourses of geology and restoration complement each other. The complex narrative structure of the novel, moreover, makes it difficult if not impossible to decide which of these discourses has precedence. On the one hand, the genre expectations raised at the start of the novel by Drendorf are destabilized, but not negated, by Risach's inner narrative. To the extent that Drendorf remains the hero of his own narrative, his geological studies take center stage. On the other hand, even if one considers Drendorf the novel's hero, Risach's life story and art-historical knowledge play a key role in the young man's education. Drendorf is on his way *back* from the mountains when he meets Risach, that is, already well on his way to becoming a geologist. Yet what Risach teaches, thus the argument of the following pages, is more than the cultural complement to the scientific life; that is, it is more than Stifter's response to the abstraction of "two cultures" (C.P. Snow) that emerged in the nineteenth century. Rather, it is the entire geological culture of this century that is expressed by the combination of these two figures and narratives: the cultural techniques of unearthing discontinuities, tracing paths, reconstructing pasts, and renouncing certain futures.

22 Cf. Schnyder, "Schrift – Bild – Sammlung – Karte," 242.

Curriculum geologicum universale

Indian Summer is not only a novel of education, but also a novel of educational programs. In the first chapter, Heinrich Drendorf describes his youth as a curriculum. Given the boy's early educational success, he receives permission from his father to pursue advanced study, at first with no other aim than becoming a "Wissenschaftler im Allgemeinen" [General Scientist].[23] This generalist attitude is both programmatic and provisional. For one, Stifter's educational experiences – from his school days in Kremsmünster to his time as a private tutor for Metternich's son – and his educational writings demonstrate the value he placed in universalist educational programs.[24] In particular, his proposed curriculum for an advanced secondary school or *Realschule* in Upper Austria *(Entwurf der Organisation einer vollständigen Realschule zu Linz für Österreich ob der Enns*, 1850) includes the entire range of modern school disciplines, from 1) German language and literature and 2) a foreign language to 3) geography and history to 4) mathematics and 5) applied arithmetic as well as 6) natural history (*Naturgeschichte*) and physics (*Naturlehre*), including 7) technology and engineering and 8) economics (*Warenkunde*) and finally 9) drawing and 10) calligraphy.[25] If the final two subjects appear dated, it is only because the media for the collection and representation of knowledge have changed and expanded. As we shall see, drawing is, in particular, a key competency in the more specialized fields that the novel explores.

And following Drendorf's career, the novel does specialize. This differentiation of knowledge begins even as Stifter evokes the *taxonomia universalis* of eighteenth-century natural history.[26] For in the movement from the general to the particular, Drendorf takes particular interest in observations of the natural world, which he divides into the animal, plant, and mineral realms. In each of these areas, he compares systematic works of natural history with his own observations: "Ich behielt die herkömmlichen Eintheilungen bei, und hatte aber auch meine Beschreibungen daneben" [I retained the conventional categories

---

23 HKG 4.1, 17 [14].

24 On the universalism of the "Mathematical Tower" in Kremsmünster, see Christian Begemann, "Metaphysik und Empirie," 92–127. On Stifter's studies, cf. Moritz Enzinger, *Adalbert Stifters Studienjahre,* and on his relationship to Metternich, cf. Wolfgang Matz, *Adalbert Stifter oder diese fürchterliche Wendung der Dinge: Biographie,* new ed. (Göttingen: Wallstein, 2016), 181.

25 Cf. HKG 10.1, 49.

26 Axel Fliethmann reads Stifter's *Indian Summer* as Foucauldian discourse analysis *avant la lettre*, cf. Fliethmann, *Stellenlektüre: Stifter, Foucault* (Tübingen: M. Niemeyer, 2001).

but also put my own descriptions beside them].[27] Although the young scientist characterizes his commentary of "conventional categories" to some extent as idiosyncratic, as "my own descriptions," the modifications and shifts of emphasis in fact reflect epochal changes in the respective fields of biology and geology. Thus in Drendorf's assessment, the traditional botanical taxonomies rely too heavily on certain visible characteristics – seed and flower parts, as Carl Linnaeus and the eighteenth century prescribed – rather than the anatomical form as a whole – as proposed in Goethe's *Versuch die Metamorphose der Pflanzen zu erklären* (*Metamorphosis of Plants,* 1790) or Alexander von Humboldt's *Ideen zu einer Physiognomik der Gewächse* (*Ideas for a Physiognomy of Plants,* 1806). Zoological observations should consider, in addition to anatomy, inherently biological qualities such as behavior, a shift in emphasis that Michel Foucault described as the difference between Buffon and Cuvier.[28] It is this very interest in the quality of life, moreover, that results in a new divide between organic and inorganic forms in the nineteenth century. Drendorf also criticizes Friedrich Mohs's system of mineralogy for its reliance on the crystallized form of minerals, which he compares to a flower's "Blühen" [blossoming].[29] This metaphor suggests that mineralogy is ultimately dependent on a botanical system of classification.[30] To mark his epoch's move toward historical geology, Stifter gives Drendorf's doubts concerning the importance of crystals an empirical foundation. In his early field studies, the young geologist finds that most minerals appear in nature in non-crystallized forms.

If Drendorf's modified classifications call into question the universal order of natural history, the course of his specialization leads to a new form of universalism which the novel calls "Erdgeschichte" [Earth History]. Attached to the global object *earth*, geology encompasses the entire natural world. In the most general sense, geology includes not only paleontology and the history of life, but also fields like meteorology that we group today under the plural heading

---

27 HKG 4.1, 32 [22]. Schneider, "Kulturerosionen," 249, suggests that this statement can serve as a model for the relationship between science (conventional categories) and literature ("my own descriptions"). If this is the case, then the uncertain knowledge of new science resides also on the side of literature.

28 HKG 4.1, 38–39 [25–26]; Foucault, *The Order of Things*, 287–304.

29 HKG 4.1, 33 [22]. Friedrich Mohs himself defines his *Charakteristik* of minerals as crystallography, "Die Kennzeichen-Lehre begreift, als ihren wesentlichen Theil, die Crystallographie," *Die Charaktere der Klassen, Ordnungen, Geschlechter und Arten oder die Charakteristik des naturhistorischen Mineral-Systems*, 2nd ed. (Dresden: Arnold, 1821) III.

30 A characterization that (again) resonates with the assessment of historians of science, see Rachel Laudan, *From Mineralogy to Geology: The Foundations of a Science, 1650-1830* (Chicago: U Chicago P, 1987), 70–86.

"earth sciences." As Risach comments, "Diese Wissenschaft ist eine sehr weite, [...] wenn sie in der Bedeutung der Erdgeschichte genommen wird. Sie schließt manche Wissenschaften ein, und sezt manche voraus" [That's a very broad science, [...] especially if it is understood in the sense of Earth History. It embraces many branches of science while presupposing a knowledge of others].[31] As we shall see, even in his excursions into art history, archeology, and restoration, the geologist remains on familiar territory, or as Risach says, "Oft habe ich gedacht, daß die Erforschung des Menschen und seines Treibens ja sogar seiner Geschichte nur ein anderer Zweig der Naturwissenschaft sei, wenn er auch für uns Menschen wichtiger ist, als er für Thiere wäre" [the study of man and his behavior, even his history, is just another branch of the natural sciences, even if it is more important for men than it would be for animals].[32]

Geology's globalism corresponds to the visual form of the panorama. During his first visits to the mountains, the young geologist hopes that a synoptic, panoramic view of the entire globe might offer a comprehensive view of its historical development. If only one could climb every mountain on earth, one would experience a series a views whose combination would reveal the earth's surface in its entirety: "wenn wir von Hochpunkt zu Hochpunkt auf unserer Erde reisen, und sie endlich alle erfüllt haben, und keine Bildung dem Auge mehr zu untersuchen bleibt als die Weite und die Wölbung des Meeres" [proceeding from high point to high point on the Earth until everything is done and you can behold no other formation that requires investigation than the broad slightly curved expanse of the ocean].[33] Both the object of this all-encompassing vision, *Bildung*, and its limit, the ocean, are significant. As "form" or "formation," the semantics of *Bildung* span discourses from geology to biology, aesthetics and education. In his critique of Goethe's *Wilhelm Meister*, Friedrich Schlegel introduced a biological concept of *Bildung* or *Bildungstrieb* (borrowed from Johann Friedrich Blumenbach) into a discussion of what would later be called the *Bildungsroman*. Goethe's novel, Schlegel argued, should be compared to an organism, whose individual parts are autonomously formed and functionally coherent, and yet through this very fact are also indicative of the whole and inseparable from it. In other words, as an organ is to the organism, so the part of the novel to the whole.[34]

---

31 HKG 4.1, 126 [76] Translation modified.

32 HKG 4.1, 122 [74].

33 HKG 4.1, 44 [29].

34 See Johannes Endres, "Über Goethes Meister," in *Friedrich Schlegel-Handbuch: Leben – Werk – Wirkung*, ed. Johannes Endres (Stuttgart: Metzler, 2017), 117–121.

The aesthetics of geological *Bildung* are similar. As we saw in the last chapter, Stifter associates the panoramic investigation of the earth with the metaphor of crystallization. The crystallized form of some minerals becomes a trope for "formative" processes more generally, including freezing and cooling. The earth's surface thus appears similar to a plate of glass on which ice crystals form:

> Wenn das Wasser in unendlich kleinen Tröpfchen, die kaum durch ein Vergrößerungsglas ersichtlich sind, aus dem Dunst der Luft sich auf die Tafeln unserer Fenster absezt, und die Kälte dazu kömmt, die nötig ist, so entsteht die Decke von Fäden Sternen Wedeln Palmen und Blumen, die wir gefrorene Fenster heißen.
>
> When moisture in the form of tiny droplets that can be scarcely seen even with a magnifying glass comes onto our window panes from the vapor in the air, and the necessary cold temperature also sets in, then the whole sheet of lines, stars, fans, palms, and blossoms that we call frosted windows is created.[35]

A similar kind of crystallization could have produced the regular shapes of mountain ranges as a hot liquid earth cooled: "[Die Erde] muß aus einem erstarrenden Stoffe entstanden sein, und streckt ihre Fächer und Palmen in großartigem Maßstabe aus" [The Earth must have originated from a hardening substance and expanded its parts to a tremendous magnitude].[36] The logic is one of microcosm to macrocosm: as on the plate of glass, so too on the large scale of the earth's surface. If this were the case, then the entire history of the earth, as subject to physical laws of crystallization, would be as visible and transparent as miniature ice crystals.

As the description of these forms as "fans, palms, and blossoms" suggests, one can imagine that crystallization is driven by a kind of organic *Bildungstrieb*. Exactly this occurs in Novalis's novel *Heinrich von Ofterdingen* (1802), an important intertext for *Indian Summer*. As we saw in the introduction, Novalis, the student of the Neptunist A.G. Werner, subscribes to the idea that the earth was once much more chaotic than it is today: "Wie ruhig und friedfertig, wie mild und klar ist gegen diese gewaltsamen, riesenmäßigen Zeiten, die heutige Natur! Und das furchtbarste Gewitter, das entsetzlichste Erdbeben in unseren Tagen ist nur ein schwacher Nachhall jener grausenvollen Geburtswehen" [How calm and peaceful, how mild and clear is today's nature when compared to these violent, gigantic times! And the most terrifying storm, the most horrible earthquake in our day

35 HKG 4.1, 43 [28].
36 HKG 4.1, 43 [28].

is only a faint echo of those dreadful birthing pains].[37] Because the earth's history has a direction, past and future become analogous, and geologists come to resemble astrologists: "Ihr [Geologen] seid beinah verkehrte Astrologen [...] Jenen ist der Himmel das Buch der Zukunft, während euch die Erde Denkmale der Urwelt zeigt" [You [geologists] are almost inverted astrologists [...] for them the heavens are the book of the future, while to you the earth reveals monuments of the primitive world].[38] The directional theory of the earth's history also allows it to be integrated into a teleological theory of *Bildung*. The metaphor of the earth's "birthing pains" relates to other geological symbols in Novalis's work. At the beginning of the novel, the protagonist Heinrich dreams of a cave. In the geological imagination of the early nineteenth century, caves often represent the primordial world both because they make stratigraphic depths accessible, and because they are often the site of spectacular fossil finds. In Heinrich's dream, the cave also has attributes of a uterus. Geology thus inspires the young poet to dream of his birth, or rather, to return to his birthplace and be reborn with a new knowledge of the past. The earth's directional development, moreover, also resembles the development of the human poet, via the intermediary of the *plant:* "[Die Natur] nähert sich dem Menschen, und wenn sie ehemals ein wildgebärender Fels war, so ist sie jetzt eine stille, treibende Pflanze, eine stumme menschliche Künstlerin" [[nature] approaches the human being, and if she was once a wild-birthing cliff, now she is a quiet, growing plant, a silent human artist].[39]

In the image of the "quiet, growing plant," Novalis invokes the symbol of the (blue) flower that runs through the novel as a whole. If the flower's blue color already suggests a certain affinity with crystals or minerals, the description of the palace garden in Klingsor's fairy tale will make this association explicit: in addition to pots "voll der mannigfaltigsten Eis- und Schneeblumen" [full of the most varied ice and snow flowers], the garden is composed "aus Metallbäumen und Kristallpflanzen" [of metal trees and crystal plants] and is covered "mit bunten Edelsteinblüten und Früchten" [with colorful gemstone blossoms and fruit].[40] This imaginary or metaphorical melding of flowers and crystals corresponds with the transitional status of Novalis's novel in the reordering of the sciences circa 1800. On the one side, eighteenth-century natural history could treat

37 Novalis, *Werke*, 201. On Werner and Novalis, see Michaela Haberkorn, *Naturhistoriker und Zeitenseher*, 194–232; on Novalis's crystallographic poetics, cf. Irene Bark, *"Steine in Potenzen": Konstruktive Rezeption der Mineralogie bei Novalis* (Tübingen: Niemeyer, 1999), 445–486.

38 Haberkorn, *Naturhistoriker und Zeitenseher*, 200.

39 Novalis, Werke, 202 [113].

40 Ibid., 233. My translation.

plants, animals, and minerals on the same discursive plane.[41] The "garden" of minerals is thus an imaginary complement to the botanical garden that demonstrates the deeper structural equivalence of botanical and mineral collections in the age of natural history. On the other side of the century's turn, Novalis's emphasis on "growth" and "formation" connects plants and minerals in a new way. Crystals undergo a process of "formation" (*Bildung*) that resembles biological growth.

Diaphanous Spaces: Mountain Lakes and Seas of Fog

By the middle of the nineteenth century, however, such analogies that reconcile the organic with the inorganic world under a unifying principle such as *Bildung* become increasingly improbable – even in literary texts that might conserve older forms of knowledge. In Stifter's imagined global panorama, the views from every mountaintop are combined such that "keine Bildung dem Auge mehr zu untersuchen bleibt" [nothing is left for the eye to investigate].[42] The figure of speech highlights the fact that the panorama is a *visual* experience, a mode of investigation or form of knowledge whose limits coincide with the limits of vision. As a smooth, opaque surface, the "ocean" presents such a limit. It is no coincidence, then, that it appears here, once again, as the poetic other to the mountain: fluid versus solid, but now also smooth versus formed.[43] The difference between smooth and ragged surfaces is also a central theme or problem in Drendorf's primary research project: the creation of a topographic map of a mountain lake. Drendorf's survey of the Lautersee closely resembles the geologist Friedrich Simony's work on Alpine lakes.[44] Because the lake's surface is opaque, the geologist must measure its depth at various points with a submerged

41 Foucault, *The Order of Things*, 136–179.

42 HKG 4.1, 44, my translation. Frye writes: "you can behold no other formation that requires investigation" [29].

43 Some of the far-reaching consequences of this distinction have notably been explored by Gilles Deleuze and Félix Guattari, *A Thousand Plateaus: Capitalism and Schizophrenia*, trans. Brian Massumi (Minneapolis: U Minnesota P, 1987). In addition to the pair smooth/striated (or stratified), the dynamics of territorialization and deterritorialization can be employed to describe the restorative efforts described in *Indian Summer:* "restoration" (by culture) is the (deterritorializing) reterritorialization of the deterritorialized (by nature and culture).

44 Banitz, "Das Geologenbild," 234; cf. Fritz Krökel, "Stifters Freundschaft mit dem Alpenforscher Friedrich Simony," *Vierteljahresschrift des Adalbert Stifter-Instituts des Landes Oberösterreichs* (1955): 97–117; and Oskar Pausch, "Neue Quellen zur Biographie Friedrich Simonys," *Jahrbuch des Adalbert Stifter-Instituts des Landes Oberösterreichs* 4 (1997): 94–121.

rope. This mode of investigation modifies the notion of surface and depth at work in the panoramic view.

In the visual study of mountain formations, the transformation of visual perception to knowledge appears immediate: forms are related to laws of formation, depth of space becomes 'depth' of time. The only limit to this visual logic is the limitation of any individual perspective. An imaginary traveling collective might "realize" all vantage points: "wenn wir von Hochpunkt zu Hochpunkt auf unserer Erde reisen, und sie endlich alle erfüllt haben" [if we proceed from high point to high point on the Earth, and they have finally all been realized].[45] In an early, programmatic discussion of his research addressed to his mountaineer assistants, Drendorf describes the "realization" of this research as the creation of maps. The map stores geological observations through a combination of image and text. While the image depicts the topography, the text consists of an index of minerals found in each location. If Mohs's taxonomy of minerals corresponds with the representational tableau of natural history, the geologist now supplants this systematic knowledge with geographical knowledge. Drendorf's index cross-references his rock collections with his maps: "Diese Stücke in ihrer Ordnung aufgestellt seien dann der Beweis dessen, was auf dem Papiere oder der Karte, wie man das Ding nenne, aufgemalt sei" [Once these are arranged in order, they will be proof of what is drawn on the paper or map – whatever you might call it].[46] This archive of geological research, with its images, maps, and samples, translates the earth into a readable visual form.[47] It is the data from which geologists infer how the mountains were formed, "aus dem, was man in den Gebirgen finde, schließen zu können, wie sie entstanden seien" [from looking at what the mountains are made of perhaps we can figure out how they came into existence].[48]

Stifter does not take the seemingly transparent reference of maps for granted, however, despite the optimism of the young geologist. A map remains after all a "thing," a piece of paper distinct from the earth itself. Drendorf's study of the mountain lake makes the artificial mediality of the map explicit. Here the visual surface that is the object of geological interpretation is hidden and must first be made visible. Indeed, topographical research creates transparency where the naked eye can no longer see. Returning to the lake for a new season of research – in a longer geological passage that will be the subject of the following pages –

45 HKG 4.1, 44; Translation modified. Frye translates "alle erfüllt haben" as "everything is done," but *erfüllen* implies a "fuller" act of realization than merely "doing" [29].
46 HKG 4.1, 232 [135].
47 Schnyder, "Schrift – Bild – Sammlung – Karte," 241–244.
48 HKG 4.1, 232 [135].

Drendorf remarks how a mountain slope is most beautiful "gegen einen Seespiegel" [beside the surface of a lake] and asks:

> Kömmt das aus Zufall, haben die abstürzenden dem See zueilenden Wässer die Berge so schön gefurcht gehöhlt geschnitten geklüftet, oder entspringt unsere Empfindung von dem Gegensatze des Wassers und der Berge, wie nehmlich das erste eine weiche glatte feine Fläche bildet, die durch die rauhen absteigenden Riffe Rinnen und Streifen geschnitten wird, während unterhalb nichts zu sehen ist, und so das Räthsel vermehrt wird?
>
> Was that a coincidence, had the rushing waters hurrying toward the lake channeled, hollowed, cut, clefted the mountains so beautifully, or did this feeling come from the contrast of water and mountains since the former is a fine smooth soft surface cut by the rough descending ledges, grooves and flutings while below nothing could be seen, thus increasing the riddle?[49]

At first, aesthetic experience leads to scientific reasoning: Can it be a coincidence? What cause is behind the mountain slope's beautiful form? As expected, the answer is geological: Falling water erodes the mountainside. The second part of the question, however, keeps the interplay of aesthetics and science in tension. The water "cuts" the mountain, but, in a *Gestalt*-reversal, one can also see how the smooth surface *is cut* by the intruding cliffs. The "Rätsel" [riddle], moreover, is complicated by the fact that the water obscures much of the mountainside. Thus, Drendorf can only *imagine* what the lake bed looks like:

> Ich dachte bei dieser Gelegenheit: wenn das Wasser durchsichtiger wäre, [...] dann müßte man das ganze innere Becken sehen, nicht so klar wie in der Luft sondern in einem grünlichen feuchten Schleier. Das müßte sehr schön sein. Ich [...] machte mehrere Messungen der Tiefe des Wassers an verschiedenen Stellen [...] Ich dachte, auf diese Weise könnte man annähernd die Gestalt des Seebeckens ergründen, könnte es zeichnen, und könnte das innere Becken von dem äußeren durch eine sanftere grünlichere Farbe unterscheiden.
>
> On this occasion I thought if the water were more transparent [...] then you would be able to see the entire inner basin, not as clearly as in the air but in a greenish watery veil. That would be exquisite. I [...] made several measurements of water depth at various points [...] I thought, in this way I could get an approximation of the lake bottom's configuration, I could draw it, distinguishing the inner part of the basin from the outer by a softer greenish color.[50]

This image too is both scientific and aesthetic. The metaphor of the semitransparent veil expresses that the scientist knows more than he can see, even as what he knows remains reliant on his vision. While the measurements are

49 HKG 4.2, 28 [190].
50 Ibid.

only an "approximation," they provide enough points for the draughtsman to fill in with lines, and indeed color, the same "softer greenish color" that he had imagined would result if the water had been more transparent. The drawing or map is thus the medium in which the imaginary panorama is given a concrete form, but at the price of abstraction.[51] The map *guides* rather than *represents*, and it is meaningful only to an educated observer. In turn, Drendorf's coloring might be read as an aesthetic compensation for this difference. The first image of the lake "in a greenish watery veil" would then re-enter the process of observation as a projection from the map. The image Drendorf calls "sehr schön" [exquisite] lies somewhere between empirical reality and drawing, in the diaphanous middle where reality is both transparent and obscure. This "diaphanous space," as Joseph Vogl writes, also qualifies the author's poetic language: "Stifters Sprache ist damit transparent und undurchsichtig zugleich. Sie beruft sich auf eine Diaphanie, in der das Wirkliche nur in seiner restlosen Übersetzbarkeit wirklich ist" [Stifer's language is thus both transparent and opaque at the same time. It invokes a diaphanous space in which the real is only real insofar as it can be translated without remainder].[52]

In the surveying of Lake Lauter – the name *lauter*, meaning "clear" or "pure," is significant – the water's initial opacity thus stands not only for the border between the visible (above) and invisible (below), but also for the border between knowledge and non-knowledge. The reversal of this relationship results in a new aesthetic-epistemic value. The surveyed lake bed appears diaphanous or semi-transparent as discovered or hypothetical knowledge that can only exist in representation. At the narrative "high point" of the novel, moreover, the chapter "The Communication," in which Risach will recount his life story – and the prehistory of Drendorf's post-estival happiness – an extended account of a winter exhibition to the "Echern" glacial plateau confirms and radicalizes this play of (in)visibility and (non)knowledge.

As Stifter's wordplay suggests – Drendorf chooses the Echern only after considering the Simmi Glacier – the episode once again fictionalizes Friedrich Simony's research, which included a winter expedition to the Dachstein glacier in 1842. The next November, in 1843, Simony recounted this adventure in a series of articles titled "Drei Dezembertage auf dem Dachsteinsgebirg" [Three December Days in the Dachstein Mountains] for the *Wiener Zeitschrift für Kunst, Liter-*

51 Cf. Albrecht Koschorke, "Das buchstabierte Panorama: Zu einer Passage in Stifters Erzählung 'Granit'," *Vierteljahresschrift des Adalbert Stifter-Instituts des Landes Oberösterreich* 38, no. 1/2 (1989): 3–13.

52 Joseph Vogl, "Der Text als Schleier: Zu Stifters *Der Nachsommer*," *Jahrbuch der deutschen Schillergesellschaft* 37 (1993), 311.

atur, Theater und Mode, the same journal that published several of Stifter's stories, including his first published work, *Der Condor* (1840).[53] As Georg Braungart has shown, both Simony's and Stifter's texts turn to the discourse of the sublime to convey the majesty of the mountain views.[54] Overwhelmed by such mountain beauty, Drendorf can thus express only his speechlessness: "Das Bild war von unbeschreiblicher Größe" [It was an image of indescribable grandeur].[55] The episode concludes, moreover, with an assurance of sublime, or indeed, transcendent fulfillment:

> Ich aber war von dem, was ich oben gesehen und gefunden hatte, vollkommen erfüllt. Die tiefe Empfindung, welche jezt immer in meinem Herzen war, und welche mich angetrieben hatte, im Winter die Höhe der Berge zu suchen, hatte mich nicht getäuscht. Ein erhabenes Gefühl war in meine Seele gekommen, fast so erhaben wie meine Liebe zu Natalien. Ja diese Liebe wurde durch das Gefühl noch gehoben und veredelt, und mit Andacht gegen Gott den Herrn, der so viel Schönes geschaffen und uns so glücklich gemacht hat, entschlief ich, als ich wieder zum ersten Male in meinem Bette in der wohnlichen Stube des Ahornhauses ruhte.

> I was brimming over with what I had seen and found. I hadn't been deceived by the profound emotion that was still in my heart which had driven me to seek the mountain heights. A sublime feeling entered my soul, almost as sublime as my love for Nathalie. Indeed, this love was even more uplifted and ennobled through this other emotion, and with a prayer to our Lord, Who had created so much beauty and granted us happiness, I fell asleep in my bed in my cozy room at Maple Inn.[56]

What could cause such elevated feelings? Although Drendorf was not "deceived" (*getäuscht*) in his expectations, the possibility of illusion and deception, of false and correct expectations, is key to this episode, which begins with the locals' surprise at Drendorf's daring proposal. When they express their fear of the extreme cold – "die Kälte sei auf den Bergen so groß, daß sie kein Mensch zu ertragen vermöge" [the cold in the mountains was so great that no one could stand it] – Drendorf responds that no one truly knows for sure what the weather is like, and that "experience" (*Erfahrung*) is better than such speculation.[57] Of course, Drendorf himself speculates about the coming weather. But he has instruments which measure *invisible* phenomena such as air pressure and humidity: "Meine

53 Friedrich Simony, "Drei Dezembertage auf dem Dachsteingebirg" in *Wiener Zeitschrift für Kunst, Literatur, Theater und Mode* 225 – 230 (November 11 – 18, 1843). Cf. Braungart, "Der Hauslehrer," 102.

54 Braungart, "Der Hauslehrer," 103.

55 HKG 4.3, 105; Frye's translation omits this sentence, cf. [379].

56 HKG 4.3, 111 [383].

57 HKG 4.3, 99 – 100.

Werkzeuge gaben starken Luftdruck und große Trockenheit an" [My instruments showed high pressure and very dry air].[58] Knowledge of the invisible remains significant for Drendorf and his hiking partner Kaspar throughout the expedition. For one thing, because of the snow cover, prior knowledge of the area is necessary for orientation: "Wir [...] hatten vor, zur Einhaltung der Richtung uns gegenseitig in unserer Kenntniß der Riffe und der Hügelgestaltungen zu unterstüzen [We [...] planned to mutually support each other with our knowledge of the ledges and hills so as to keep the right direction].[59] At one point, Drendorf even turns to his compass for reassurance. On their return, though, the two hikers can follow their own footsteps in the snow, that is, the physical traces of the knowledge they have acquired. With knowledge of invisible meteorological forces, moreover, Drendorf not only chooses the best moment for the expedition in advance, but he can also interpret the fog that forms below the mountain heights and calm his companion:

> Kaspar wollte erschrecken. Ich aber machte ihn aufmerksam, daß der Himmel über uns ganz heiter sei, daß dieser Nebel von jenem sehr verschieden sei, der [...] den Wanderern so fürchterlich ist; unser Nebel sei kein Hochnebel sondern ein Tiefnebel, der die Bergspizen, auf denen das Verirren so schrecklich sei, freilasse, und der beim Höhersteigen der Sonne verschwinden werde.
>
> Kasper became alarmed. However, I pointed out to him that the sky above us was perfectly clear and that this fog was different from the fog that [...] was dangerous to hikers. However, our fog wasn't up on the heights – it was in the lower levels, and the peaks where getting lost is so fearful were left clear. This fog would disappear as the sun rose higher in the sky.[60]

Although it frightens Kaspar, the opaque fog is as much a source of aesthetic delight in this episode as anything else. Like the snow and the ice, the fog stands in for the beauty and sublimity of the *sea*. It is no coincidence that Drendorf's companion is named Kaspar. The image of "indescribable grandeur" closely resembles one of the most recognizable works of the most famous German Romantic painter, Caspar David Friedrich, "Wanderer über dem Nebelmeer" [Wanderer above the Sea of Fog, ca. 1818]. Here is Stifter's description of the "sea of fog" that Drendorf and Kaspar encounter:

> Hier hatten wir einen eigenthümlichen Anblick. [...] Die Wirkung, welche sich aus dem Aneinandergrenzen der oberen wärmeren Luft und der unteren kälteren, wie ich schon

58 HKG 4.3, 101. My translation of an omitted sentence, [377].
59 HKG 4.3, 103 [378].
60 HKG 4.3, 107–108 [381].

> am schwarzen Steine bemerkt hatte, ergab, war noch stärker geworden, und ein einfaches wagerechtes weißlichgraues Nebelmeer war zu meinen Füssen ausgespannt. Es schien riesig groß zu sein, und ich über ihm in der Luft zu schweben. Einzelne schwarze Knollen von Felsen ragten über dasselbe empor, dann dehnte es sich weithin, ein trübblauer Strich entfernter Gebirge zog an seinem Rande, und dann war der gesättigte goldgelbe ganz reine Himmel, an dem eine grelle fast strahlenlose Sonne stand, zu ihrem Untergange bereit.
>
> Here we had a remarkable view. [...] The effect of the higher warmer air bordering on the layer of cold below such as I had already noticed at the black rock became more pronounced, resulting in a simple horizontal whitish gray sea of fog spread out at my feet. It seemed colossal, it was as if I were floating in the air above it. Individual black knobs of cliff towered above it; it extended great distances and was bordered on the horizon by a dull blue strip of distant mountains. The totally clear sky was saturated with yellow gold from the dazzling rayless sun just about to set.[61]

The "black knobs of cliff," in particular, recall Friedrich's painting, which Stifter enhances with the image of the "rayless sun," even as "saturated with yellow gold" gives a fairly close approximation the painter's hazy sky. Stifter's phrase "floating on air" translates well the noble pose of Friedrich's wanderer, who, though we do not see his face, must be experiencing the same elevated feelings reported by Stifter's protagonist.

Beauty through Knowledge: The "Winter Letters from Kirchschlag"

The sea of fog also has a scientific explanation, "the effect of the higher warmer air bordering on the layer of cold below," which points to invisible or unimaginable phenomena, without translating it into the aesthetic feeling of the sublime. The visual indicator of temperature difference is the fog itself, which acts like a thermometer for the expert observer. The reality described by meteorological discourse, however, cannot be seen or depicted. Stifter explores this difference between depiction and explanation extensively in a later text, which, next to Simony's Dachstein expedition and Friedrich's painting, can be seen as a third point of comparison for Drendorf's winter mountain ascent. In the popular science essay "Winterbriefe aus Kirchschlag" [Winter Letters from Kirchschlag, 1866], Stifter divides the two modes of representation into distinct letters: The first five *explain* certain foundations of physics, while the sixth attempts to *depict* the "beauty" of the natural spectacle produced by these physical laws.

61 HKG 4.3, 105 [379].

As the essay title suggests, the letter-writing author assumes the perspective of a man writing from the top of a mountain, in this case, at Kirchschlag near Linz. His aim is to explain why he has sought out the mountain as a cure for his ailments, even if "Meinung" [opinion] and "Brauch" [custom] forbid such visits in winter.[62] To this end, he must "zuerst im Allgemeinen von einigen Dingen [...] reden, welche die Grundwirkungen auf das menschliche Leben ausüben, und welche auf Höhen anders sind als in Niederungen" [first speak in general about several things which exert fundamental effects on human life and which differ by altitude].[63] These "things," which Stifter also calls "höchste Lebensdinge" [the highest things of life] and "Lebensgrundlage" [the foundation of life] comprise *light, warmth, electricity, air,* and *water.*[64] Each of the first five letters explores one of these topics, which makes the "Winter Letters" a valuable source for Stifter's scientific knowledge and interests, at least in the 1860s. The observation, for example, that altitude often translates into warmer temperatures in the winter months, finds its place in the second letter, on warmth. The letter goes on to explain, moreover, how this temperature difference causes the formation of fog: "so scharf grenzen oft warme und kalte Luft an einander, daß die Oberfläche des Nebels von oben gesehen meistens eine wagrechte Fläche wie eine riesige ebene Tischplatte bildet" [often the warm and cold air form such a sharp border that the surface of the fog, seen from above, forms a horizontal plane like a gigantic flat tabletop].[65]

In the sixth letter, Stifter relates this sober, scientific description of the "sea of fog" to the aesthetic feelings that it produces. For Stifter, cloud and fog formations are "external," and the feeling of beauty is an "internal" thing: "Da ich nun bezüglich des Lebens und Gedeihens der Menschen auf dem Kirchschlagberge über verschiedene äußerliche Dinge gesprochen habe, will ich auch von innerlichen reden, die nicht minderen, vielleicht noch größeren Einfluß auf unser Dasein nehmen" [since I have now spoken about various external things concerning the life and thriving of people on the Kirchschlag mountain, I want now to speak of the internal things which have no less, and perhaps an even greater influence on our being].[66] Here Stifter proposes a theory for the relationship between aesthetics and science. In essence, the feeling of beauty develops and is elevated by knowledge, in a process of increasing intensification: "Je mehr das Gefühl für Schönheit angeboren und durch sich und durch Kenntnisse entwick-

62 Adalbert Stifter, "Winterbriefe aus Kirchschlag," PRA 15, 257. Translations are my own.
63 PRA 15, 258.
64 PRA 15, 260; 266.
65 PRA 15, 264.
66 PRA 15, 279.

elt ist, desto höhere empfinden wir, und desto höher empfinden wir sie" [The more the feeling for beauty is given at birth and is developed through itself and through knowledge, the higher is the beauty we feel, and the stronger we feel it].[67] His first example is the majesty of the starry sky, which only becomes greater with improved knowledge of astronomical facts, e.g., that light, which races at 40,000 miles a second, requires millennia to reach us from many stars; that a million earths would fit within the sun; or that the fog of the Milky Way contains a multitude of solar systems. The mountaintop, as Stifter writes, is the perfect location for unimpeded views of all the universe has to offer: the stars, but also sunrise and sunset; the atmosphere in its blue color and spectacle of clouds, as well as the forms of the mountains and valleys. At the same time, the letter repeatedly compares the sublime mountains with the sublime ocean: "Außer dem Meere habe ich nie etwas Schöneres auf der Erde gesehen" [Other than the ocean, I have never see anything more beautiful on earth].[68] Indeed, when one morning, the weather leads to an unexpected storm with a tornado funnel, the author can say, "Und der Anblick des Schauspiels dieses Morgens war für mich noch erhabener als der des Meeres" [And the view of the spectacle this morning was for me even more sublime than that of the ocean].[69]

Stifter's invocations of the discourse of the sublime have a twofold structure. First, the sublime is a form of the beautiful that elevates us because it fills us with knowledge of nature's grandeur. That is, scientific knowledge – astronomical facts, meteorological theories – takes the place of the ideas of reason in the Kantian schema. Knowledge elevates beauty and makes it "higher" or more sublime. Stifter is also attracted, however, to the terrifying moment of the sublime experience. This is the second form of the sublime in Stifter. In such instances, it is the *lack* of knowledge of natural forces that threatens the subject's integrity. This sublime, which Michael Minden has compared with the "uncanny" or the "postmodern sublime," results from Stifter's implicit rejection of the Kantian idealist paradigm.[70] Whereas Kant can distinguish between the "mathematical" sublime, which is resolved by the ideas of pure reason, and the "dynamic" sublime, assuaged by moral ideas of practical reason, Stifter's reduction of all ideas to scientific knowledge precludes such a division of labor. Instead, knowable phenomena become elevatingly beautiful, whereas unknown, or unknowable

67 Ibid.

68 PRA 15, 282.

69 PRA 15, 283.

70 Michael Minden, "Stifter and the Postmodern Sublime," in *History, Text, Value*, ed. Minden, Swales, and Weiss-Sussex, 9–21.

phenomena are simply terrifying: the formation of a tornado during a storm is thus "more sublime," not "more beautiful."

Drendorf's self-reflections on his chosen discipline of earth history bear witness to this tension. As we have seen, the discussion of the survey of Lake Lauter demonstrates how a topographic reconstruction can create the illusion of depth beneath a beautiful surface. In this case, knowledge enhances and elevates the aesthetic feeling. Yet as that central passage continues, it demonstrates the destabilizing potential of geological knowledge. As Peter Schnyder notes, this passage has a remarkable formal characteristic: the present tense in the geologist's questions interrupts the preterite (past historic) of his autobiographical narrative.[71] In addition to underlying the significance of this passage, the present tense emphasizes the present action or "actuality" of geological forces. In other words, it demonstrates the corollary of Charles Lyell's famous dictum summarizing his actualism: "The present is the key to the past." If present geophysics can be used methodologically to understand the geological past, then the same forces of the past, *which created and moved mountains and oceans*, must be acting in the present. The geological present thus engulfs Drendorf's account of his own, remarkably short human life. His autobiography makes use of a past tense that has no meaning within the overarching present of geological change. On a larger scale, moreover, the history of the earth engulfs all of human history. At the end of the passage, Drendorf will call human history a mere interpolation or "Einschiebsel."

The motif of the lake plays a central role in the geological convulsion that shakes up Drendorf's worldview. His survey of the lake bed leads him to far-reaching questions: "Diese Bestrebungen brachten mich auf die Betrachtung der Seltsamkeiten unserer Erdgestaltungen" [These efforts brought me to a consideration of the peculiarities of earth formations].[72] Among these "peculiarities" is the complementary relationship between erosion and sedimentation. The material eroded from the mountainside falls into the lake, which slowly fills up over "in Jahrtausenden und Jahrtausenden [...] bis einmal, mögen hundert oder noch mehr Jahrtausende vergangen sein, kein See mehr ist" [many thousands of years

---

71 Cf. Schnyder, "Schrift – Bild – Sammlung – Karte," 239. Rejecting a commonplace among Stifter scholars, Schyder concludes: "Wie sich gezeigt hat, hat Heinrichs Interesse an der Geologie nichts mit einer Flucht aus der Zeitlichkeit, nichts mit einer Flucht in eine 'heilsame Starre' zu tun" [As has become clear, Heinrich's interest in geology has nothing to do with a flight from temporality, nothing to do with a flight into a 'healing stasis'], 244.

72 HGK 4.2, 29 [190].

[...] until sometime, be it in a hundred thousand or more years, there is no more lake].[73] Given enough time, land can become sea and sea land.

But there is more: the land is itself fluid. After considering the phenomenon of sedimentation, Drendorf observes the "slope" ("Neigung") of the strata as evidence of the rising of the mountains.[74] This theory of *Erhebung*, which "elevates" the earth, not the soul, is precisely what disturbed Goethe and made him so reluctant to accept the new evidence of Vulcanism discovered by Alexander von Humboldt and Leopold von Buch, among others. By contrast, Stifter's geologist not only asks whether the mountains are still rising, but also indulges in speculations that resemble future theories of plate tectonics:

> Hört die Hebungskraft einmal auf? Ist nach Jahrmillionen die Erde weiter abgekühlt, ist ihre Rinde dicker, so daß der heiße Fluß in ihrem Innern seine Kristalle nicht mehr durch sie empor zu treiben vermag? Oder legt er langsam und unmerklich stets die Ränder dieser Rinde auseinander, wenn er durch sie seine Geschiebe hinan hebt?
>
> Will this elevating force ever cease? Has the Earth cooled more after millions of years, is its crust thicker so that the hot stream in its core is no longer able to force its crystals up? Or is it slowly and inconspicuously eating away at the edges of this crust when it pushes stones (*Geschiebe*) through it?[75]

The self-confidence with which Novalis met the pacification of geological change is clearly gone. In addition to shifting to the present tense, Stifter employs a rhetoric of paratactic amplification to underscore Drendorf's shock as he reflects on one geological phenomenon after another: erosion followed by sedimentation, elevation followed by erratic blocks ("Fremdlinge") and fossils, then elevation once again and glaciation, weathering and erosion, sedimentation, elevation, climate change and elevation.[76] This list reflects both a survey of established geological knowledge circa 1850 – such as was well known to Stifter from his study of Baumgartner's earlier *Naturlehre*, for example – as well as a report on most contemporary research, for example the work of Friedrich Simony.

Perhaps most astonishing for readers in the twenty-first century is the attention that Stifter pays to various forms of climate change: "Werden Winde Wolken Regen nicht anders?" [Won't the winds, clouds and rain become different?][77] The

---

73 HKG 4.2, 29 [190].
74 HKG 4.2, 29 [191].
75 HKG 4.2, 32 [192].
76 Cf. HKG 4.2, 30. On erratic blocks in Stifter, cf. Jason Groves, "Erratische Blöcke: Stifters 'Bunte Steine,'" in *Sema: Wendepunkte der Philologie*, ed. Joachim Harst and Kristina Mendicino (Würzburg: Königshausen & Neumann, 2013), 255–269.
77 HKG 4.2, 32 [192].

insight into the changing surface of the earth is so radical that it leads Drendorf to wonder whether the agents of this change are malleable as well. Such speculation also correlates with the particular concern that Stifter shows throughout his work – and systematically in the "Winter Letters from Kirchschlag" – for the living milieu and conditions of life. Thus the filling-in of the lake bed fascinates in particular because it creates "ein Thal, an dem jezt die Ufer als grüne Wälle in langen Strecken stehen, mit kräftigen Kräutern blühenden Büschen und mancher lachenden Wohnung von Menschen prangen" [a valley with luxuriant green walls of bushes and growth along with many a cheerful dwelling of man].[78] Or in the opposite case, a forest of petrified wood is particularly disturbing because it is evidence of the disappearance of these conditions: "Sind die Berge gestiegen, und haben sie ihren Wälderschmuck in höhere todbringende Lüfte gehoben? Oder hat sich der Boden geändert, oder waren die Gletscherverhältnisse andere?" [Did the mountains rise and bring their crowning glory of forest into the more rarified, deadlier air? Or had the ground changed, or were the glacial conditions different?].[79]

In addition to revealing a contemporary understanding of radical geological change, Drendorf's reflections also have important semiotic and poetological implications. As Albrecht Koschorke demonstrated in his reading of the novella *Granit*, Stifter's realism often relies on cartographic space to serve as a stable reference of literary signs. Such signs refer to the real, in other words, to the extent that they "spell out a panorama."[80] If the ground can shift, however, if a lake can become a valley or a valley a mountain, then this referent loses its basis in reality.[81]

The poetological solution to this problem is narrative. One tells a story about how the landscape has changed. While slow-moving, Stifter's narratives complement the apparently synchronic semiotics of his descriptions with the diachrony of deep time. Indeed, by the end of the passage, Drendorf's geological thinking takes an emphatic *narrative* or *historical* turn: "Wenn eine Geschichte des Nachdenkens und Forschenswerth ist, so ist es die Geschichte der Erde, die ahnungsreichste, die reizendste, die es gibt [...]" [If any history is worth pondering, worth investigating, it is the history of the Earth, the most promising, the most stimulating history there is].[82] As the indefinite article, "any history," indicates, *Ge-*

78 HKG 4.2, 29 [191].
79 HKG 4.2, 31 [191–192].
80 Cf. Koschorke, "Das buchstabierte Panorama," 3–13.
81 This radicalizes the instability of cultural signs (monuments or gravestones) discussed in Chapter 1.
82 HKG 4.2, 32 [191].

schichte appears here in its double meaning of "story" and "history." Unlike natural history or *Naturgeschichte*, earth history is a history in the sense of a narrative account of past events. As discussed in the introduction, Martin Rudwick proposes the term *geohistory* in order to distinguish this historical approach to the earth's past from earlier geological theory.[83] The redefinition of geology as geohistory implies the radical temporalization of nature and humanity's position within it that we now call the discovery of "deep time."

Narratives and histories, however, also have a medial form. In the case of earth history, the medium is the stone itself:

> Die Quellen zur Geschichte der Erde bewahrt sie selber wie in einem Schriftengewölbe in ihrem Innern auf, Quellen, die vielleicht in Millionen Urkunden niedergelegt sind, und bei denen es nur darauf ankömmt, daß wir sie lesen lernen, und sie durch Eifer und Rechthaberei nicht verfälschen. Wer wird diese Geschichte einmal klar vor Augen haben? Wird eine solche Zeit kommen, oder wird sie nur der immer ganz wissen, der sie von Ewigkeit her gewußt hat?
>
> The Earth itself preserves the sources (*Quellen*) of this history in its innermost parts just as in a room for records (*Schriftengewölbe*), sources inscribed in perhaps millions of documents (*Urkunden*); it is only a matter of our learning to read and not falsify them by eagerness or obstinacy. Who will one day have this history clearly before his eyes? Will ever such a time come, or will only He know it completely Who has known it for all Eternity?[84]

The earth's core is a "room for records" with "millions of documents." While the image of a vault of writings in the earth's interior is marked with the "as" of simile, it also traverses the metaphorical field in the manner of an absolute metaphor.[85] Thus, the word "Gewölbe" both evokes a vaulted storage "room" that recalls the catacombs and the "Red Stone" from *Castle Crazy*, and the "vaults" of the earth's interior, the varied arcs of its contours: caves, ravines, mountains.[86] Likewise, the "sources" (*Quellen*) of the earth's history are "like" historical documents (*Urkunden*) – but what is a source here? It is surely not literally a spring, a fount in the earth that one might expect if the difference between geology and

83 Other histories of geology make similar arguments, cf. Roy Porter, *The Making of Geology: Earth Science in Britain, 1660–1815*, and Rachel Laudan, *From Mineralogy to Geology: The Foundations of a Science, 1650-1830*. As the subtitles of the last two studies suggest, the dates 1815–1830 designate an approximate time frame for geology's "establishment."

84 HKG 4.2, 32–33 [192].

85 Hans Blumenberg, *Paradigmen zu einer Metaphorologie*, 2nd ed. (Frankfurt a.M.: Suhrkamp, 1999). See also the discussion of Goethe in the introduction.

86 In *Die Narrenburg*, the autobiographical writings of the Scharnasts are stored in "Der rothe Stein," a part of the castle complex and a section of the novella, "in einem düstern, weiten Gewölbe [...] wie in einer Gruft," HKG 1.1, 383.

history could be upheld. Once again, Stifter's text unearths a metaphor, the source, and places it in or returns it to its presumably original context in the realm of nature. But the description of the context "earth" also requires metaphors: crucially, the notion of earth *history* already contains a metaphor that implies both its legibility and the existence of "sources," or legible signs. The literary text refreshes the metaphors of "source" and "vault," so to speak, but this reactualization only extenuates the tension between the literal and figural meanings of words.

In suggesting that this history is one that we might learn to read, this passage evokes the traditional metaphor of the "book of nature" discussed in the introduction. According to Hans Blumenberg, we should recall, the metaphor of nature as book traditionally has two motivations. First, it represents the competition between science and Scripture, between human knowledge of the "book of nature" and the divine revelation contained in the "book of books." Second, the image of the book lends unity and coherency to nature.[87] The unity of the bound volume suggests that nature, while vast, may ultimately be limited and comprehensible. Stifter's text thus marks both an endpoint for the metaphor of the book of nature and a new beginning in the history of the world's legibility. The plurality of "millions of documents" distinguishes the metaphor of the "room of records" (*Schriftengewölbe*) from the earlier unifying image of the "book of nature." Various sources may be deciphered without necessarily forming an ordered totality. Just as history books are never complete because the future is open and history does not end, so too are earth history books.

Earth history thus represents a form of the sublime – "the most promising, the most stimulating history" (*ahnungsreichste* suggests "promise" but also "foreboding") – associated with a purely hypothetical idea of totality: "Who will one day have this history clearly before his eyes?" It is no longer sufficient, as Goethe imagined in the essay "On Granite," to "survey" a mountain range in order to survey the earth's history.[88] As a "room of records," the earth resembles less a map or a book than a monstrous or labyrinthine *library*.[89] In a library, stories and histories only appear in the plural. And thus earth history: "eine Geschichte, in welcher die der Menschen nur ein Einschiebsel ist, und wer weiß es, welch ein kleines, da sie von anderen Geschichten vielleicht höherer Wesen abgelöset werden kann" [a history where man is only an interpolation,

87 Hans Blumenberg, *Die Lesbarkeit der Welt*, 17–18.

88 Goethe, "Granit II," 12; "On Granite (1784)," 914. See Introduction.

89 On the monstrosity of the library, cf. Nikolaus Wegmann, *Bücherlabyrinthe: Suchen und Finden im alexandrinischen Zeitalter* (Köln: Böhlau, 2000), 47–77.

who knows how small a one, and can be superseded by other histories of perhaps higher beings].[90] The word Stifter uses to describe the history of humanity, *Einschiebsel*, which means "interpolation" or "insert," suggests not only, as the text explicates, that this history might be unimaginably brief and insignificant, one likely to be superseded by greater or longer histories. Borrowed from the context of printing, the word is also a metaphor that extends the metaphor of writing at work in this passage. Unlike "book" or even "room of records," the "interpolation" denotes an addition, inserted after printing, a supplement, or even something incongruent, a meaningless excursion or digression.

At the same time, the choice of a word that includes the root *schieben*, "to push," in a discussion of geology, is doubly meaningful. Rubble or detritus from erosion is called *Geschiebe*, the rocks that flowing water has washed away to a new place. With this play of words, Stifter closes this important passage with a return to its beginning: "Als ich in dem Frühling die Hauptstadt verlassen hatte, [...] war ich einmal bei einem Haufen von Geschiebe stehen geblieben, das man aus einem Flußbette genommen, und an der Straße aufgeschüttet hatte, und hatte das Ding gleichsam mit Ehrfurcht betrachtet" [In the spring when I had left our capital [...] I once stopped by a pile of stones they had taken from a river bed and strewn by the road. I regarded this with a sense of awe].[91] A "pile of stones" (*Haufen von Geschiebe*, also a "heap of rubble") leads Drendorf to reflect on the implications that earth history has for the earth-historical position of humanity, but this pile also has a significant position at the edges of Vienna, the capital city and modern metropolis. The rubble presumably arrived in the riverbed as a result of natural forces, but as the geologist notes, the force that brought it up onto the street was *cultural*. While Drendorf recognizes the difference between the human and natural forces that shaped and transported these rocks, he nonetheless connects these two types of action in an extended analogy:

> Ich erkannte in den rothen weißen grauen schwarzgelben und gesprenkelten Steinen, welche lauter plattgerundete Gestalten hatten, die Bothen von unserem Gebirge, ich erkannte jeden aus seiner Felsenstadt, von der er sich losgetrennt hatte, und von der er ausgesendet worden war. Hier lag er unter Kameraden, deren Geburtstätte oft viele Meilen von der seinigen entfernt ist, alle waren sie an Gestalt gleich geworden, und alle harrten, daß sie zerschlagen und zu der Straße verwendet würden.
>
> In the red, white, gray, dark yellow and speckled stones all with flat rounded shapes, I saw the messengers of our mountains; I could tell the craggy home of each one, where it had

90 HKG 4.2, 32 [192].
91 HKG 4.2, 27 [190].

> been separated and from whence it had been sent out. Here it lay among comrades whose birthplaces were often many miles distant from its own; all of them had assumed the same shape, and all were waiting to be broken up and used on the road.[92]

Here the rocks are "messengers" (*Bothen*), each with its own *Felsenstadt:* a "craggy home" or more literally, "cliff city." Like the "room of records," the message that they convey is the story of their origin and subsequent displacements. But what will the stones have to say about their new mission, to be crushed and employed to construct a road? The use of the stones in the construction of a human city erases whatever past the stones might have had in their "cliff cities." As we saw in the first chapter, Stifter identifies in "A Walk Through the Catacombs" smooth roads with intensified modern communication and the social and cultural indifference it produces. Here, the road that provides access to the mountains also threatens to efface the mountainous past of its material. Drendorf's questions, "wozu dann alles da sei, wie es entstanden sei, wie es zusammenhänge, und wie es zu unserem Herzen spreche" [why is all that here, how did it come to be, what is it made of and what does it tell us], applies as much to the street as it does to the stones, to history as it does to geology.[93] This brief passage already answers two questions that concern the fate of the stones; and while the geologist can explain how the stones arrived at the riverbed, only the historian can explain how they were removed.

The Poetics of Restoration

Represented in this analogy by the riverbank, the disciplinary boundary between geology and history is porous and susceptible to shifting. In one direction, road-builders can always lift stones out of the riverbed and into the realm of history and culture. In the other direction, the geologist can always see in the material basis for historical and cultural life – architecture and the world of "things" in general – a past that ascends far back beyond the beginnings of this culture. Drendorf turns away from his geological reflections and "flees" to the poets not only because literature promises to give meaning to his scientific endeavors.[94] His perspective on his older friend Risach's cultural conservation efforts

---

92 HKG 4.2, 27–28 [190]. Translation modified for clarity; Frye translates "Bothen" as "harbingers."

93 HKG 4.2, 28 [190].

94 "Von solchen Fragen flüchtete ich zu den Dichtern" [from such questions I fled to the writers], HKG 4.2, 33 [193].

also uncovers their epistemic ground. Stifter's novel is thus constructed through and through to demonstrate the reciprocal interplay of geo-historical science and cultural-historical meaning.

As suggested above, if the model for Heinrich Drendorf is Friedrich Simony, then the model for the Freiherr von Risach is the author Stifter himself. Like the author of "A Walk Through the Catacombs" and "The Flea Market," the amateur art historian and conservator Risach is aware of the latest scientific developments of his time. From the start, his conversations with the young geologist demonstrate knowledge of the field. Indeed, in the friendly meteorological wager that sets the plot of the novel in motion, it is Risach who correctly predicts that there will be no thunderstorm. In addition to meteorological observation and study of contemporary scientific works such as Alexander von Humboldt's *Voyage aux Régions équinoxiales du Nouveau Continent, fait en 1799, 1800, 1801, 1802, 1803 et 1804* (*Personal Narrative of Travels to the Equinoctial Regions of America During the Years 1799–1804*),[95] Risach also has remarkable foresight about the future effects of the modern scientific and technical revolution on the horizon:

> [W]ir stehen noch zu sehr in dem Brausen dieses Anfanges, um die Ergebnisse beurtheilen, ja wir stehen erst ganz am Anfange des Anfanges. Wie wird es sein, wenn wir mit der Schnelligkeit des Blizes Nachrichten über die ganze Erde werden verbreiten können, wenn wir selber mit großer Geschwindigkeit und in kurzer Zeit an die verschiedensten Stellen der Erde werden gelangen, und wenn wir mit gleicher Schnelligkeit große Lasten werden befördern können? Werden die Güter der Erde da nicht durch die Möglichkeit des leichten Austauschens gemeinsam werden, daß allen Alles zugänglich ist? [...] Dann wird, um der Allberührung genügen zu können, das, was der Geringste wissen und können muß, um Vieles größer sein als jetzt.
>
> We are still too much in the confusion of the initial stages to be able to assess the results; in fact, we're only in the initial stages of the very beginning. How will it be when man can send news over the whole world with the speed of light as well as travelling at great speeds to the most remote areas, not to mention transporting great burdens equally fast? Won't the goods of the Earth be shared more because of the ease of exchange? Won't everything be accessible to everyone? [...] Then in order to meet the challenge of all this contact, even the lowliest will have to know much more than they do now.[96]

The technological and industrial globalization that Risach envisions here is the cultural complement of Drendorf's global geological speculation. Both men pay

---

95 Cf. Schößler, "Der Weltreisende Alexander von Humboldt in den österreichischen Bergen," 261–285.

96 HKG 4.2, 227–228 [300–301].

particular attention to the material side of global change. In addition to the electrical communication of "lightning," Risach considers the communication of heavier materials, "große Lasten," which necessitates heavier infrastructure. In other words, this passage translates the cultural criticism of the younger Stifter of *Vienna and the Viennese* into a more optimistic or indeed, utopian vision of global exchange of knowledge and goods. This global connectedness (*Allberührung*) is both the result of the scientific research undertaken by generations of scientists like Drendorf, and the condition of the ultimate knowledge that he seeks.

Risach's faith in scientific progress and his optimistic view of technology are hardly exceptional in the context of nineteenth-century culture, though such views are often forgotten in discussions of Stifter's "restorative" poetics. For Stifter, the recuperation of the past often occurs in the service of the future. This is at least the function of "restoration" or *Wiederherstellung* as Risach describes it. Admittedly, the initial motivation of conservation work is the "charm" (*Reiz*) of antiquities:

> Hier werden Dinge, [...] welche lange vor uns ja oft mehrere Jahrhunderte vor unserer Zeit verfertigt worden, und in Verfall gerathen sind, wieder hergestellt, wenigstens so weit es die Zeit und die Umstände nur immer erlauben. Es wohnt in den alten Geräthen beinahe wie in alten Bildern ein Reiz des Vergangenen und Abgeblühten, der bei dem Menschen, wenn er in die höheren Jahre kömmt, immer stärker wird. Darum sucht er das zu erhalten, was der Vergangenheit angehört, wie er ja auch eine Vergangenheit hat, die nicht mehr recht zu der frischen Gegenwart der rings um ihn Aufwachsenden paßt. Darum haben wir hier eine Anstalt für Geräthe des Altertums gegründet, die wir dem Untergange entreißen zusammenstellen reinigen glätten und wieder in die Wohnlichkeit einzuführen suchen.
>
> Here things that were made a long time ago – sometimes even centuries before our own time – and have fallen into decay are refinished and repaired; they are redone at least as much as time and circumstances allow. In the old pieces of furniture almost as in the old paintings, dwells the charm (*Reiz*) of the past, of the faded which becomes stronger in a man when he reaches an advanced age. Thus, he seeks to preserve what belongs to the past just as he too has a past that isn't quite suitable to the fresh and alive present, the growing and flourishing that surrounds him. So we founded a little place for old pieces of furniture which we save from destruction, put together again, clean, polish, and make suitable for use in a household.[97]

Reiz is a key term in Stifter's writings, which he uses as a synonym for beauty, but particularly with respect to its phenomenal or physical manifestation. The term also appears when Stifter expresses the sentiment that art or beauty is

97 HKG 4.1, 96 [59].

the terrestrial manifestation of the divine: "das höchste irdische Gut ist die Kunst, die Darstellung des Göttlichen im Kleide des Reizes dargestellt" [the highest earthly good is art, the representation of the divine in a dress of charm].[98] Antiquities have a special *Reiz* because they bear the signs of age, the "charm of the past, of the faded." In addition to this aesthetic motivation, Risach adds a personal one: As he becomes older, the appeal of old things grows. Conservation (*Erhaltung*) becomes a form of self-conservation (*Selbsterhaltung*). This comment foreshadows Risach's later revelation that the entire "house of roses" is a restoration of his own past – or rather, a past he wishes he had lived. Even in its short, proleptic form, however, the notion of conserving the things that are part of one's own life, yet already in the historical past, implies the loss at the heart of the conservation effort: the "destruction" (*Untergang*) can be delayed, but not completely avoided.

Autobiography of a "Conservator"

The informal circle of men, "welche auf die Dinge des Alterthums ihr Augenmerk richteten" [interested in things of the past], and which Drendorf already describes as an "Anstalt" [institution], positions Risach's conservation work in the context of Stifter's own personal and historical past.[99] If we take seriously the dating of the novel's fictional events to the 1820s or 1830s, then this first step at the institutionalization of historical conservation may point to Anton Ritter von Spaun's founding, in 1833, of the *Verein des vaterländischen Museums für Österreich ob der Enns* [*Association of the National Museum for Austria above the Enns*].[100] Stifter deeply admired the literary and cultural historian Spaun, whose speculations on the authorship of the *Nibelungenlied* the author integrated into his novel *Witiko*.[101] As is fitting for the genre of historical fiction, Risach describes the next step of institutionalization in the future tense: "Es wird einmal eine Zeit kommen, in welcher vom Staate aus vollkommen sachverständige Männer in ein Amt werden vereinigt werden, das die Wiederherstellung alter

98 HKG 8.1, 52. On the Jesuit context of this topos, see Karl Möseneder, "Schriften zur Kunst," in *Stifter-Handbuch*, ed. Begemann and Guiriato, 162. On Stifter's theory of art, cf. Begemann, *Die Welt der Zeichen*, 359–411, who deconstructs the Platonism of the "dress" metaphor.

99 HKG 4.1, 98 [60].

100 E. Lebensaft, "Spaun, Anton von (1790–1849)," *Österreichisches Biographisches Lexikon 1815–1950*, vol. 13 (Lfg. 59), (Vienna: Österreichische Akademie der Wissenschaften, 2007), 5–6.

101 Cf. Cornelia Blasberg, *Erschriebene Tradition*, 317–321. For further discussion, see Chapter 4.

Kunstwerke einleiten, ihre Aufstellung in dem ursprünglichen Sinne bewirken, und ihre Verunstaltung für kommende Zeiten verhindert wird" [Someday there will come a time when the state will bring experts together in one department who will propose the restoration of valuable old works of art, effect this in keeping with the original spirit, and prevent their being marred or damaged in subsequent times].[102]

This time came in 1853, with the founding of the *K.K. Centralkommision zur Erhaltung der Baudenkmäler in Österreich* [*Royal-Imperial Central Commision for the Conservation of Architectural Monuments in Austria*]. Just days after the official start of the commission in January of this year, Stifter wrote a letter to Antonie Arneth (née Adamberger), the wife of "Regierungsrat" Joseph Arneth, who had been given the task of leading the Commission: "Ihren hochverehrten Gatten bitte ich zu grüßen, und ihm zu sagen, daß ich ihn beneide, daß er sich nur mit lauter Schönem zu beschäftigen braucht, und jezt wieder in der Section für Erhaltung vaterländischer Baudenkmale beschäftigt ist. Eine solche Stellung würde mich außerordentlich freuen" [Please pass along my greetings to your most honored husband and tell him that I envy him that he only needs to deal with such beautiful things, and is now once again employed in the Section for the Conservation of National Architectural Monuments. I would be extremely happy to have such a position].[103] Thanks to his longstanding friendship with Antonie and Joseph Arneth, and the restoration work he had already begun on a wooden altar from Kefermarkt, Stifter's wish soon came true: on December 15, he was officially named *Conservator* for Upper Austria.

Stifter could thus rightly consider himself one of the "experts" whose duty it is to conserve and restore old works of art. Accordingly, the restoration projects described in his novel draw from his own work as a conservator, from the Kefermarkt altar, which appears as a wooden altar found in "Kerberg," to the unrealized plan "zum Ausbaue des schönen Thurms der schönen gothischen Kirche zu Braunau" [to complete the beautiful tower of the beautiful gothic church in Braunau] – the town's name means something like "Brownfield" and hence becomes in the novel the fictional "Grünau" or "Greenfield."[104] In describing such projects in the novel, moreover, Stifter employs the same specific language as in his writings on conservation. In an article for the Carolinum Museum in Linz "Ueber den geschnitzen Hochalter zu Kefermarkt" [On the Carved High Alter in Kefermarkt], for example, Stifter writes: "es ist eine der ersten Pflichten eines Volkes, das sich

102 HKG 4.1, 112 [68].

103 PRA 18, 146f., cited in Wilfried Lipp, "Stifter als 'Conservator,'" 186.

104 On Kefermerkt/Kerberg, cf. HKG 4.2, 142 [253]; cited in Lipp, "Stifter als 'Conservator,'" 189, and "Grünau,"cf. HKG 4.1, 113 [69].

selber achtet, seine Kunstdenkmäler auch zu achten, sie zu schützen, zu erhalten und dieselben, wenn sie durch Zeit und Barbarei gelitten hätten, wieder, so weit es möglich ist, in den ursprünglichen Stand zu setzen" [It is one of the first duties of a people, which respects itself, to respect its art monuments as well, to protect and conserve them, and to restore them, if they have suffered from time and barbarism, as far as possible to their original condition].[105] Respecting this duty, the Representative (*Statthalter*) for Upper Austria Eduard Freiherr von Bach made the decision, upon viewing the Kefermarkt Altar in 1852, in Stifter's account, "eine Rettung des Werkes vor äusserstem Verfalle und eine Wiederherstellung in dem ursprünglichen Sinne einzuleiten" [to begin a salvaging of the work from the most extreme dilapidation and a restoration in the original spirit].[106]

In his assessment of such statements in Stifter's conservation reports, Wilfried Lipp concludes that the Upper Austrian's attitude towards restoration can be summed up as an adherence to the "doctrine de l'unité de style," the leading international trend of his day. The French phrasing of this doctrine of stylistic unity associates it with Viollet-le-Duc, whose theories are explored in greater detail in the next chapter. In the history of conservationism, including Walter Frobl's history of conservation in Austria, which informs Lipp's analysis, this doctrine of restoration comes under heavy fire from critics who see it as a naïve denial of the reality of historical change, or a "retro-restoration that eliminates the effects of time."[107] In this history, which largely parallels the history of editorial practices beginning with Lachmann and his *Urtext*, the naïve notion of the "original" state was thankfully overcome in favor of notions like the "age value" (*Alterswert:* Alois Riegel) or Georg Dehio's motto "conservation, *not* restoration."[108] As Lipp himself shows, however, early conservators were aware of the problematic described by the distinction conserve/restore. In the guidelines of the Royal-Imperial Commission, for example, one reads: "[Die Restaurationen] werden sich [...] auf die Ergänzung solcher Theile ausdehnen, durch deren Mangel ein weiterer Verfall die Folge ist. Sie haben sich aber nicht auf die Ergänzung abhängiger, in den Charakter oder den Baustyl eingreifender Bestandtheile zu erstrecken, selbst wenn eine solche Ergänzung in dem Geiste der Ueberreste vorzunehmen beabsichtiget würde" [Restorations should extend to the completion

105 HKG 8.4, 80.

106 HKG 8.4, 82.

107 "Zurückrestaurieren unter Tilgung des Faktors Zeit," Lipp, "Stifter als 'Conservator,'" 191. Cf. Walter Frodl, *Idee und Verwirklichung. Das Werden der staatlichen Denkmalpflege in Österreich* (Vienna: Böhlau, 1988).

108 Cf. Georg Dehio and Alois Riegel, *Konservieren, nicht restaurieren. Streitschriften zur Denkmalpflege um 1900*, ed. Marion Wohlleben and Georg Mörsch (Braunschweig: Vieweg, 1988).

of those parts which lacking would be the cause of further dilapidation. They should not extend to the completion of dependent components that affect the character or the building style, even if such a completion may be intended in the spirit of the remains].[109] In other words, and much in the spirit of today's more "conservative" architectural restorations, structurally necessary components should be restored, but not components that are presumed to "fit" the original aesthetic spirit.

Both the "conservation" and the "restoration" positions were thus well established in the conservation discourse circa 1850.[110] Although conservation practices have developed in the direction of more limited interventions, the problem of defining what exactly constitutes despair or dilapidation (*Verfall*) remains. For those, like Stifter, who tend to favor restoration, a major source of degradation of historical monuments comes from the renovations of more recent periods. The Kefermarkt Altar, for example, has been subject to "abscheuliche Verbesserungen" [despicable improvements] and has "durch Zeit und Barbarei viel gelitten" [suffered greatly from time and barbarism].[111] Barbaric refurbishment also ruined the façade of the "Star Court" or "Sternenhof" in *Indian Summer,* a façade which the palace's owner Mathilde, under Risach's counsel, decides to remove. With the perspective of historical distance, we can observe how the most virulent polemics against an earlier generation of restorators repeat the same critique of the recent past. Max Dvorák, for example, complains in 1911 about the era of historicist restorations as a "Periode, der weit mehr Denkmäler zum Opfer gefallen sind, als je durch Kriege und Revolutionen vernichtet wurden" [period that many more monuments fell vicitim to than had ever been destroyed by wars and revolutions].[112]

Whether or not one intervenes to repair "barbaric" historical damage – be it from political or aesthetic revolutions – one needs to know what constitutes a historical epoch in the first place. As Stifter's depictions of conservation work in *Indian Summer* make clear, not only is the "unity of style" an answer to this problem, it is an answer that *only* the nineteenth century could give, that is, after the advent of historical geology and "geological" history or archeology. The narrative perspective of the novel is an important device in communicating

109 Quoted in Lipp, "Stifter als 'Conservator,'" 187; cf. Frobl, *Idee und Verwirklichung*, 192–204.
110 Cf. Norbert Huse, *Denkmalpflege: deutsche Texte aus drei Jahrhunderten* (Munich: Beck, 1984), 84–123.
111 HKG 8.4, 69; 80.
112 Max Dvorák, "Denkmalpflege in Österreich [1911]," *Österreichische Zeitschrift für Kunst und Denkmalpflege* 28, no. 3 (1974): 131f. Quoted in Lipp, "Stifter als 'Conservator,'" 191.

this discursive connection: a geologist (Drendorf) reports what he learns from the conservator (Risach).

Most striking is the similarity between geological topography and the drawings of the conservator – which the conservator Lipp even today calls, without explicit analogy to the geosciences, "kunsttopographische Erhebungen und Aufzeichnungen" [art-topographical surveys and drawings].[113] As we saw above, Drendorf explains the importance of topography for geological research. The geological map reproduces the earth-historical archive, transcribing and translating it into a more readable form. The Freiherr von Risach's practice of art-historical research follows the same method. The archive of this research is also twofold. First, Risach presumes the existence of an archive of Art: *die Kunst*, a universal collection of all art-historical objects that resembles the "geological record" or "room of records" in its totality. Second, he collects a comprehensive archive of drawings that aspires, like the geological map, to reproduce this universal archive. With the help of these drawings, Risach can hope not only to identify the historical epoch of the pieces in his art collection, but also to use this knowledge of historical styles "die Wiederherstellung alter Kunstwerke ein[zu]leiten, ihre Aufstellung in dem ursprünglichen Sinne" [to propose the restoration of valuable old worlds of art, effect this in keeping with the original spirit].[114]

After studying the Freiherr von Risach's large collection of architectural drawings, the young geologist Drendorf has an idea remarkably similar to the organicism of contemporary architects like Viollet-le-Duc:

> Da ich nun alle diese Laubwerke diese Ranken diese Zacken diese Schwingungen diese Schnecken in großer Abfolge sah, erschienen sie mir gewissermassen wie Naturdinge etwa wie eine Pflanzenwelt mit ihren zugehörigen Thieren. Ich dachte, man könnte sie eben so zu einem Gegenstande der Betrachtung und der Forschung machen wie die wirklichen Pflanzen und andere Hervorbringungen der Erde, wenn sie hier auch nur eine steinerne Welt sind. Ich hatte das nie recht beachtet, wenn ich auch hin und wieder an einer Kirche oder an einem anderen Gebäude einen steinernen Stengel oder eine Rose oder eine Distelspize oder einem Säulenschaft oder die Vergitterung einer Thür ansah. Ich nahm mir vor, diese Gegenstände nun genauer zu beobachten.
>
> As I now looked at all this foliage, these tendrils, points, curved lines, spirals in great succession (*in großer Abfolge*), they seemed to me like natural objects, a plant kingdom with its appropriate animals. I thought: a person could make them the subject of observation and research just as well as the real plants and other things of the Earth, even if they are only a world in stone (*eine steinerne Welt*). I had never really paid attention when now and again I

113 Lipp, "Stifter als 'Conservator,'" 186.
114 HKG 4.1, 112 [69].

saw a stone stem, a rose, a thistle top or a column on a church or other building or the metal grillwork on a door. I resolved to take closer notice of these things in the future.[115]

Once again, Stifter employs here a phrase that points to a deep discursive connection through understated ambiguity: "a world of stone" is, after all, what the geologist studies as well as the historian of architecture. And though the novel's conservator, the Freiherr von Risach does not identify comparative anatomy as an inspiration, he does operate according to the same principles. Risach explains his method:

> Diese Dinge haben so gut wie bedeutendere Gegenstände ihre Geschichte, und aus dieser Geschichte kann man das Aussehen und den Bau derselben zusammen sezen. Im Verlaufe der Jahre habe sich die Gestaltungen der Geräte immer neu abgelöset, und wenn man auf diese Abfolge sein Augenmerk richtet, so kann man aus dem vorhandenen Ganzen auf verloren gegangene Theile schließen, und aus aufgefundenen Theilen auf das Ganze gelangen.

> These things, like more significant objects, have their own history, and from this you can deduce the appearance and construction. In the course of the years the shape and forms changed predictably, and if you pay close attention to the sequences (*Abfolge*), from an available whole you can deduce lost pieces and reconstruct the whole from what is extant.[116]

The things in question here are furnishings, which, in the first instance, follow the same principles of development as all other, "more significant" art objects, including buildings. Yet important here is also the relationship of furniture to architecture, "[...] daß die Bauwerke einer Zeit und die Geräte, welche in diesen Bauwerken sein sollten, eine Einheit bilden, die nicht zerrissen werden kann" [that buildings and the appropriate furniture of a given period form a unity that should not be torn asunder].[117] Changing forms result from changing "styles," and the true expression of style is an architectural ensemble, interior and exterior, a *milieu.* Though Risach's own museum-like house more resembles the eclectic historicism typical of bourgeois residences, he aids his friend Mathilde in furnishing her palace, the *Sternenhof* or "Star Court," according to this principle: each room is the reconstructed milieu of a particular epoch in the history of the palace.[118]

---

115 HKG 4.1, 104 [63–64].

116 HKG 4.1, 98 [60].

117 HKG 4.1, 108 [65].

118 Kristin Belgum has thus read the novel's interiors in the context of aesthetic historicism or the "high historicism" of the 1850s. Subsuming her analysis under the notion of "seamlessness,"

Restoration in Practice: Marble Figures

As the next chapter will show, this theory closely resembles Viollet-le-Duc's organicist and evolutionary theories of art history and restoration. The examples of restoration work in *Indian Summer* demonstrate how complicated the process of reconstruction can become in practice. In addition to the restoration projects based on Stifter's own work (the Kefermarkt altar and the plan to restore a church tower in Braunau), these examples include an antique marble statue of a nymph, a Renaissance painting of the Virgin with Child, and a set of medieval woodcarved wall panels. The distribution of these projects along the highpoints of Western art history is programmatic. If *Indian Summer* pays homage to contemporary enthusiasm for the Gothic, evidenced by the awe Drendorf feels in front of the "Kerberg" altar, its horizon is more universal. Risach considers the art of all ages, from all continents, and imagines art history within the expanded horizon of geological deep time, the only scale large enough to encompass the beginning and end of humanity: "Es wäre des höchsten Wunsches würdig, wenn nach Abschluß des Menschlichen ein Geist die gesammte Kunst des menschlichen Geschlechtes von ihrem Entstehen bis zu ihrem Vergehen zusammenfassen und überschauen dürfte" [It would fulfill my highest desire if after we leave this moral sphere our spirit could survey and embrace the entire artistic expression of the human race from its beginnings to its end].[119]

As his companion Mathilde affectionately jokes, "Das wäre ja im Großen, was du jezt im Kleinen thust, und es dürfte hiezu eine ewige Zeit und ein unendlicher Raum nöthig sein" [that would be doing on a large scale what you are now doing on a small, and for that scale an infinite time and endless space would be necessary].[120] The universalist perspective of a *longue durée* frames further aspects of Risach's conservation work that are connected to geology. While the heap of rubble and the analogy of the "cliff city" (*Felsenstadt*) hints at an intermingling of geology and history, the overlapping of culture and nature finds its symbol in marble. As Drendorf exclaims, "wie bewunderungswürdig ist der Marmor! Wo sind denn die Thiere hin, deren Spuren wir ahnungsvoll in diesen Gebilden sehen? Seit welcher Zeit sind die Riesenschnecken verschwunden, deren Andenken uns hier überliefert wird?" [how remarkable marble is! Where did the

however, she overlooks the differences between the Rose House and the Star Court, cf. Belgum, "High Historicism and Narrative Restoration," *The Germanic Review* 67, no. 1 (1992): 20.

119 HKG 4.2, 146 [255].

120 Ibid.

animals go whose traces we think we see in these formations? When did the giant snails disappear whose memory has been handed down to us here?][121]

Given the significance of marble in Drendorf's geological reflections, it should not be surprising that it holds a privileged position in the architecture of the Rose House. The main entrance to the Rose House leads directly to a hallway paved with marble. The official tour of the house begins here, continues up a marble staircase, past the antique marble statue, and into a hall that contains nothing but a collection of marble tiles, artfully arranged to cover the floor and walls. At the center of this hall, moreover, are red marble Ammonites: the "giant snails" that both resemble works of art and recall the deep geological past.[122] Lightning strikes at critical moments in the novel's plot, and always illuminates marble surfaces. It is a marble statue that teaches the young geologist Drendorf to appreciate beauty. Having asked Risach why he was not told before that the statue is beautiful, the mentor assures him that it is better to learn by oneself to see beauty. The statue is, moreover, a transitional object in an erotic sense.[123] The sight leads Drendorf first to a symbolization of desire, a recollection of Homer's Nausicaa then to a more appropriate "object" of desire, Natalie, to whom he declares his love with yet another marble statue as witness.[124] The marble hall of the Rose House will then serve as the location where the couple officially marries.

Marble symbolizes the convergence of geology and art history because it can be conceived of as both rock matter and artwork. As artwork, it recalls Greek antiquity, the origin and highest epoch of classic or philhellenic art history. As rock matter, however, marble belongs to the world of geology. The Rose House embodies marble's double significance. While the entrance, staircase and statue reflect the importance of marble as a sign of art in its highest forms from the an-

---

121 HKG 4.2, 30 [191].

122 As Schnyder has discovered, such Ammonites, called *Ammonites Metternichii*, were not only part of Friedrich Simony's collection in Hallstatt, but they also were featured in Alphonse von Morlot's *Erläuterungen zur geologischen Übersichtskarte der nordöstlichen Alpen* (1847) a book that Stifter owned, cf. Erwin Streitfeld, "Aus Adalbert Stifters Bibliothek. Nach den Bücher- und Handschriften-Verzeichnissen in den Verlassenschaftsakten von Adalbert und Amalie Stifter," *Jahrbuch der Raabe-Gesellschaft* 18 (2010): 142. Morlot's description includes a prominent analogy between natural (fossils) and cultural objects (coins). Cf. Schnyder, "Schrift – Bild – Sammlung – Karte," 246.

123 The scene in which Drendorf first notices the beauty of the statue is appropriately called "Die Annäherung."

124 Nausicaa is mentioned at HKG 4.2, 74 [215]. Drendorf declares his love at HKG 4.2, 251–269 [314–324]; The "fountain nymph," 314, at the Sternenhof is more recent in age than the one at the Rose House, just as Natalie is younger than Mathilde.

cient temple to the reception halls of modern palaces, the marble hall showcases the value of the marble itself as an object of collection, admiration, and study. Taken together, the marbled rooms of Risach's house constitute a temple of art *and* nature.

Furthermore, as a material, marble has a double aspect. Later in the novel, when Natalie asks Heinrich if he considers "den bloßen Stein schon für sehr schön" [the stone itself to be very beautiful], the young geologist answers predictably: "Ich halte ihn für ganz besonders schön" [I consider it particularly beautiful].[125] She then identifies two elements of marble's beauty: its depth and, paradoxically, its transparency. To explain this beauty, Drendorf employs a metaphor similar to the ice crystals metaphor used to represent the formation of mountains:

> "Der Stein ist auch durchsichtig," antwortete ich, "nur muß man eine dünne Schichte haben, durch die man sehen will. Dann scheint die Welt fast goldartig, wenn man sie durch ihn ansieht. Wenn mehrere Schichten übereinander liegen, so werden sie in ihrem Anblicke von außen weiß, wie der Schnee, der auch aus lauter durchsichtigen kleinen Eisnadeln besteht, weiß wird, wenn Millionen solcher Nadeln auf einander liegen."
>
> "But the stone is transparent," I answered, "you just have to have a thin layer to look through. If you look through this layer the whole world seems almost golden. When several layers are on top of each other, they seem white, just as snow, which consists of nothing but transparent tiny needles, also seems white when millions and millions of these needles on top of each other."[126]

Just as the mountain ranges resemble the ice crystals that form on a glass surface, so the crystalline marble resembles the crystals that make up snow. What is more, this metaphor suggests that the beauty of the world stems from its composition in layers. The geological gaze, that is, the gaze that identifies layers, can see the world both through the "thin layer" as "almost golden," and as a composite of layers that demonstrate "eine sehr große Tiefe, als sollte man in ihn eindringen können" [a very great depth there, as if you should be able to penetrate into it], the first quality of marble to which Natalie attests, in addition to the appearance of transparency, "und als wäre er durchsichtig, was er nicht ist" [but also as if it were transparent, which it isn't].[127] In this way, the marble represents another diaphanous space like the lake mirror and the sea of fog discussed

125 HKG 4.2, 254 [316].
126 HKG 4.2, 254–255 [316].
127 Ibid.

above. Through this combination of transparency and depth, marble thus serves as a microcosm or synecdoche for the geologist's earth as a whole.

If the meeting of Heinrich and Natalie in front of the marble fountain nymph is a highpoint in the novel's plot, the most spectacular description of a marble statue, in the entrance hall of the Rose House, occurs earlier. The passage begins: "eines Abends, da Blize fast um den ganzen Gesichtskreis leuchteten, und ich von dem Garten gegen das Haus ging, fand ich die Thür, welche zu dem Gange des Amonitenmarmors zu der breiten Marmortreppe und zu dem Marmorsaale führte, offen stehen" [One evening when lightning was flashing almost over the entire horizon and I was coming from the garden toward the house, I found the door leading to the ammonite marble corridor, the broad marble stairway and the marble hall standing open].[128] The lightning not only signals the momentousness of the occasion, but also reveals the geniality of Risach's architectural design. As Drendorf climbs the marble staircase – in felt slippers, so as not to damage the marble – he is transfixed by the figure on the landing, illuminated by occasional bursts of lightning through the glass ceiling above. The temple Stifter envisions here could hardly be more modern. The glass ceiling, a recent architectural innovation in the mid-nineteenth century, has since become a ubiquitous feature of museum architecture.[129] The purpose of this installation is to create a sacred space through architectural technique. Unlike wall windows, "durch die alltägliche Gegenstände herein schauen, und durch die verworrene Lichter einströmen" [where ordinary things could gaze in, where a confusion of light would be streaming in], the glass ceiling communicates directly with the light of the heavens.[130] The image of refracted light as "confusion" contrasts in particular with the topos of lightning as a message from God. The text is quite clear, moreover, in noting that this is a temple *of* light, not a temple *in which* light happens to shine. The light here surrounds the statue "mit einer dämmerigen Helle wie mit einem Tempel" [with a twilight brightness as if with a temple].[131] The lightning itself produces a temporary

128 HKG 4.2, 72 [214].

129 On Stifter and contemporary architecture, see Arburg, "Neues von der *Narrenburg?*" 109–133. On museum aesthetics, or "mausoleum" aesthetics, cf. Glaser, *Die Restauration des Schönen*, 227. For a more nuanced reading of the "museum environment" in the novel, see Katharina Grätz, who emphasizes active processes of "living tradition" and "creation of meaning" in Stifter's work, *Musealer Historismus*, 214–244; cf. also Silke Arnold-De Simine, "Musealisierungsphänomene im Werk Adalbert Stifters," in *Ordnung-Raum-Ritual*, ed. Becker and Grätz, 41–68.

130 HKG 4.2, 74–75 [215].

131 HKG 4.2, 75. My translation: Stifter's phrasing is so idiosyncratic here that Frye simplifies by translating "surrounding it with the light of the temple" [215].

"rosenrotes Licht" [rose-red light] on the white marble, an effect that gives the statue the illusion of life: "Ich hatte eine Empfindung, als ob ich bei einem lebenden schweigenden Wesen stände, und hatte fast einen Schauer, als ob sich das Mädchen in jedem Augenblicke regen würde" [I had a feeling as if I were standing beside a silent living being and almost shuddered to think that the maiden might move at any moment].[132] This is Ovid's Pygmalion, Eichendorff's *Marmorbild*, but technically produced.[133] It engages the fortuity of a natural spectacle, lightning, to reproduce the practice, widespread in the early nineteenth century, of viewing antique marbles at night by torchlight. Classicist art theory thus ascribes to Greek art the unique ability to produce an illusion of life. At the same time, the metaphorics of Renaissance and revival suggest that the goal of classicism should be to revive this ancient ability. Compared to Eichendorff, however, Stifter's description of the resurrection or *Wiederbelebung* of the antique statue is sober and technical.[134] The text notes every aspect of the artifice that produces a sensation before the sensation itself is named.

Drendorf soon moves on to the marble hall, where he meets the Freiherr von Risach pacing in contemplation. The marble hall contains Risach's collection of marbles, though it is no ordinary scientific collection. For one, despite a large variety of Alpine marble, it contains only samples that Risach has collected himself: it is a collector's collection.[135] And though Risach's interest in marble may

132 Frye's "pink light" effaces the symbolic "rose" in Stifter's diction, ibid.

133 The novel plays an interesting game with its intertexts. This passage refers directly to Homer: "Ich dachte an Nausikae, wie sie an der Pforte des goldenen Saales stand, und zu Odysseus die Worte sagte: 'Fremdling, wenn du in dein Land kömmst, so gedenke meiner" [I thought of Nausicaa as she stood at the portal of the golden hall saying to Odysseus, 'Stranger, if you reach your homeland, remember me'] HKG 4.2 74 [215]. Nonetheless, the allusions to the Ovidian tradition are unmistakable, and clearly structure the scene itself. Stifter's technique is itself restorative: in the place of modern (sentimental, decadent, Romantic) texts that clearly underlie his poetic construction, he inserts a more pure, classical tradition: Homer, Shakespeare, Cervantes, Goethe. On the "traces" of Greek antiquity in the novel, which produce a conflict between "timeless" antiquity and the time of history, see also Gabriella Catalano, "Griechische Spuren in Stifters *Nachsommer*," in *Graecomania: Der europäische Philhellenismus*, ed. Gilbert Heß, Elena Agazzi, and Elisabeth Décultot (Berlin: De Gruyter, 2009), 99–115.

134 As Franziska Schößler notes, Stifter suppresses both the eroticism and *Grauen* that characterize Eichendorff's text, "Rahmen, Hüllen, Kleider und das Phantasma der Durchsichtigkeit. Verschwindende Medien in Stifters *Nachsommer*," in: Tina-Karen Pusse (ed.) *Rhetoriken des Verschwindens* (Würzburg: Königshausen & Neumann, 2008), 110.

135 In this way, it bridges the gap between the principle of completeness that dominates in scientific collections, and of representativeness in art collections, cf. Katharina Grätz, "Musealisierung und Sammlung," 307. It is not possible, as Grätz suggests, to collect "all species of marble;"

be in part scientific, he puts his collection to an artistic purpose by using it as decoration. The hall, moreover, has a remarkable architectural feature: a series of windows grants a panoramic view of the mountains. During the crepuscular lightning storm, this view shows itself in all its sublime grandeur. Menacing clouds rise solemnly over the distant Alps, creating a sublime spectacle: "Zu dem Ernst der Wolkenwände gesellt sich der Ernste der Wände von Marmor" [The gravity (*Ernst*) of the marble walls combined with the majesty (*Ernst*) of the towering thunderheads].[136] Here again, the lightning demonstrates its effect, "der reine kalte Marmor [war] wie in eine Glut getaucht" [a few times the pure cold marble seemed bathed in fire].[137]

The installation of the marble hall is as artificial as that of the marble statue, yet the juxtaposition of the two goes beyond the parallel spectacles of art and science, the beautiful and the sublime. The description of the marble hall in this passage is relatively short, but significant in terms of narrative structure. It occurs during a pause in the conversation between Drendorf and Risach about the marble statue and its restoration: "Er schwieg nach diesen Worten, und ich sah den Mann an. Wir waren, während er sprach, in dem Saale auf und nieder gegangen" [After this he was silent, and I gazed at him. While he was speaking, we walked up and down the hall].[138] The marble hall is a setting; in light of the place of geology in the novel, however, it is better understood as a foundation. Stifter's "installation" of geology at the center of his novel, below and around the discussions of art, lays the groundwork for a theory of art in which Art itself – as a pure abstraction – may remain eternal, but in which art works are temporal and historical, destined for death despite all efforts of restoration and revitalization (*Wiederbelebung*).

As we saw above, Stifter's depictions of the sublime mountain panorama are accompanied by scientific explanations of "how it all came to be." Similarly, Risach explains in great detail how the ancient Greek statue came to his house in Upper Austria – indeed, he eventually explains how everything in his house, and the house itself, came to be. As Saskia Haag has shown in an insightful analysis of the novel's restoration projects, restoration always involves transportation and movement.[139] In the case of the Greek marble, it once stood in a "Bretterbude bei Cumä in Italien" [shed in Cuma, Italy]. Because the square adjoining the tavern

---

types of rock are specific to their locality, or as we have seen, they are "messengers" from their "cliff cities."

136 HKG 4.2, 86 [221].

137 HKG 4.2, 86 [221–222].

138 HKG 4.2, 86 [221].

139 Haag, "Versetzt. Restaurierung als Entortung," 77–87.

was used for ball games, the statue was eventually covered up to the bust with wood.[140] Other details indicate the various forms of "barbarism" under which the statue suffered: surrounded by other plaster sculptures of lesser worth, it is put up for sale with these because the tavern owner's son has a plan to replace them all with new plasters, whose equal size indicates modern mass production: "die schönsten Gipsarbeiten, [...] die alle gleich groß waren" [the most beautiful plaster statues, all the same size].[141]

In Risach's story, the transport of the statue back to Austria is a decisive factor in revealing its worth. The conservator narrates the process of discovery step by step in a way that parallels Drendorf's at times pedantic descriptions of his geological research. The first such scene has voyeuristic undertones: "die Besizerin, eine wahrhaftige altrömische Sibille [...] führte mich zu dem Mädchen, und zeigte mir durch eine Spalte der Bretter, daß dasselbe in ganzer Gestalt da stehe, und also die anderen Dinge weit überrage" [the owner, a genuine Ancient Roman Sybille [...] took me to the figure of the girl and pointed out through the cracks in the boards that it was a full statue and thus would tower far above the other things].[142] On Risach's second visit to the tavern, the statue has been freed of its wooden base, and moved to a shed containing building materials. Here Risach can view "die früher nicht gesehene übrige Gestalt" [the rest of the figure of the maiden which I hadn't seen earlier], and which he judges as "wesentlich nicht verletzt" [not appreciably damaged] and aesthetically "außerordentlich" [extraordinary].[143] Having now bought the statue, he has it packed again carefully for transport. A revealing clue as to the true nature of the statue is the high price of the freight, which Risach confirms upon inspecting the package. As this too is removed and the statue is placed in a well-lit restoration studio, where the contours of the statue, all its "Schwingungen und Schwellungen der Gestaltung" [ridges and swellings of the form] become visible.[144] It is not until the restoration work begins, with a meticulous dusting and careful cleaning of the statue's surface, which culminates in light scraping with a knife and file, that the chief restorator Eustach comes across a spot on the shoulder that appears to be marble. Further examination of the heel show that there too, a layer of plaster easily breaks of to reveal the marble beneath.

Stifter's description of all these – and indeed – other steps in the transport and restoration of the statue is long and exhaustive. At various moments, from

140 HKG 4.2, 76 [216].
141 HKG 4.2, 77 [217].
142 HKG 4.2, 77–78 [217].
143 HKG 4.2, 78 [217].
144 HKG 4.2, 80 [218].

the first voyeuristic peek to the last, spot-by-spot examination of the statue's plaster-covered body, the passage emphasizes the cupidinous desire of the collector's gaze. Yet Risach returns to Italy to pay the tavern owner for the difference in value, and receives in return a notarized bill of sale for the ancient, marble object, a noble gesture that corresponds with Stifter's earlier criticism of antiquarian greed in the essay "The Flea Market." The value of the work has thus been recovered in a double sense: aesthetically restored and economically restituted. With the notarized exchange, the transport of the work has also been "signed." In the logic of the earth's "room of records," or of the geologist Drendorf's topographic maps, this signature guarantees the possibility of return, even if this final act of restitution remains hypothetical.

Risach also assiduously researches the history of a Renaissance painting found by his apprentice Roland in an attic. If the locus of this find is significant in the context of Stifter's work – for example, the fictional manuscript of *My Great-Grandfather's Notebook* was also found in an attic – and indeed in the wider antiquarian imagination of the nineteenth century, its content and historical epoch fill an important gap in Risach's collection, and complement Hellenistic paganism with Christianity, antiquity with modernity. This restoration also provides a variant on the theme of transformation of value via transport. In this case, the canvas itself was used to pack a soldier's laundry. Of the two moves that Risach's research is able to discover, then, one causes the complete devaluation of the work and the other the complete restoration of this value. Needless to say, the owner of the attic is also duly compensated for the unexpected find.

Frames and Environments

In addition to the economic restitution that accompanies the restoration, the example of the Renaissance painting leads into extended reflection on the relationship between painting and frame. This passage begins with an analogy that asserts the value of the frame: "Es ist nicht wahr, was man öfter sagt, daß eine schöne Frau ohne Schmuck schöner sei als in demselben: und eben so ist es nicht wahr, daß ein Gemälde zu seiner Geltung nicht des Rahmens bedürfe. Ich hatte zu unserem Marienbilde einen Rahmen nach Zeichnungen aus mittelalterlichen Gegenständen bestellt" [It isn't true what they often say about a beautiful woman without jewelry being more beautiful than with jewelry; likewise it isn't true that a painting doesn't need a frame for maximum effect].[145] If this

145 HKG 4.2, 108 [234].

analogy to jewelry suggests that the framer is an artist, or at least artisan, then it also implies that restoration is akin to frame-making. Both framing and restoring are operations of contextualization or re-situation. Risach makes a distinction, however, between contemporary framing and the belated framing of a restored work of art. When the painter and the framer are contemporaries (or indeed the same artist), then the painter will often modify her painting once she has seen it in the frame. In other words, there is an accommodation between painting and frame that goes in both directions, and which justifies the view that the painting is only truly a complete artwork within its frame. The particular problem of restorative framing comes from the fact that the restorer cannot change the original painting: "Wie dann das Bild in einem Rahmen aussehen werde, liegt nicht in der Willkühr des Wiederherstellers, und wenn es in dem Rahmen trefflich oder minder gut steht, so ist das Sache des ursprünglichen Meisters, dessen Werk man nicht ändern darf" [How the painting will look in a frame doesn't depend on the arbitrary decision of the restorer; whether it is beautiful or less suitable in the frame is the responsibility of the original master whose work you may not change].[146] The syntax of this sentence reflects the complexity of a train of thought that appears to contradict itself. It follows, however, from Stifter's starting-point, which may stem from his own practice as a painter, namely that the "normal" case is a collaboration between painter and framer. When this case is not given, then the original work must have the particular power to look good in "the frame" (a generalizing definite pronoun), that is, to look "beautiful or less suitable" in any number of *different* frames. The masterwork must be able "sich selber durch seine Gewalt geltend [zu] mach[en]" [to dominate its frame] if it is to be ready for its historical fate – and astonishingly, the works of the past have shown this sort of resistance and adaptability.[147]

The choice of frame, of course, is in fact an "arbitrary decision of the restorer." In this case, Risach chooses a "strong" (*mächtig*) frame, full of decorative elements in a high medieval style. There is no certainty about the period of the painting, but the speculations all point to the Renaissance: Guido Reni, Titian, or the Rafael School.[148] In the context of a theory of restoration based on the notion of the "unity of style," this disjuncture of frame and painting is surprising. It may reflect an attempt to establish historical plausibility where historical authenticity is lacking. After all, there is no reason why a Renaissance painting might not have been placed originally in an older, late medieval frame.

146 HKG 4.2, 109 [234].
147 Ibid.
148 HKG 4.2, 110 [235].

Indeed, this anachronism repeats the curious timeline of Stifter's example in which a painting is only finished *after* the frame has been chosen. The frame appears to have priority. In any case, however, Risach's frame choice insists on the frame as a *choice* of the restorer. The anachronism takes the form of Viollet-le-Duc's famous definition of restoration as the re-establishment of a state of completion that may never before have existed.

In Haag's analysis of restoration as a transfer or transport operation, the reconstruction of architectural elements such as the medieval wooden wall panels poses a similar problem. Such decorative architecture not only forms a "unity of style" like all other "mobile" furniture, but also a material unity with the "immobile" building in which they were first installed: *Möbel* matches *Immobilien*. The problem of recreating a room to fit these panels proves unresolvable, not despite the restoration efforts, but *as a result of them*. Because Drendorf only finds some of the original panels, his father initiates a limited restoration that fills in certain gaps and positions them in an extension of their home in Vienna. Upon its completion, he declares, "Was aber von weit größerer Bedeutung ist, besteht darin, daß das Holzkunstwerk in das ganze Häuschen so paßt, als wäre sie [!] ursprünglich für dasselbe bestimmt gewesen – und dies freut mich am meisten" [Of far greater importance is the fact that this wooden work of art harmonizes with the entire house as if it [/she] had originally been intended for here – and that most of all makes me happy].[149] The original German contains a revealing grammatical error – the "it" should read the neuter *es* for the closer antecedent "das Holzkunstwerk," rather than the feminine "sie," which can only refer back to "die Holzverkleidung." The slip exposes the denial at work in the unreal assumption of "as if," the suspension of disbelief in this aesthetic fiction: it is "as if" the panels are a "Verkleidung," a covering or dress made to fit the walls. But the panels were *not* originally destined for this space, nor can any restoration that resituates them achieve this state of completeness. After all, they were only ever brought there because they are a "work of art," which, in the best of cases, will "dominate" the framing space.

The restoration does not end there, however. As a surprise for his young friend, Risach undertakes, on his own account, the reconstruction of the missing panels. Once these are added to the original panels, the ensemble no longer fits in the space that Drendorf's father had created for them. As a result, the elder Drendorf decides to reinstall the now "completed" restoration in the Drenhof, a villa that he has purchased for his newly married son. The re-installation, though, exceeds the diegesis of the novel. If this story demonstrates the transi-

149 HKG 4.2, 127–128 [244].

tory character of all restorations, which must remain as incomplete as the incomplete original, it also shows how one can move forward in an imperfect and transitory world. Drendorf's novel of education still follows the classic logic of progressive development or perfectibility: he surpasses his father and thus exceeds the space of his paternal home. As a mentor, Risach is a surrogate for the young man's father, but one who also allows him to go beyond the limits of his origins.

Ultimately, Risach's restoration work, and the novel as a whole, is thus oriented towards the future. As we saw above, Risach believes in cultural progress, at least in the scientific and technological realms. Not only does he think that art historical knowledge will increase as more objects are "dug up," but also that conservation work can lead to a revival of modern artistic practices. In addition to his restorations, the artisan Eustach makes furniture that may be inspired by older forms and techniques, but does not imitate them. Likewise, his brother Roland, a painter, helps restore old masterworks, but also paints in an emphatically contemporary style. It is no accident, finally, that Roland's paintings are *geological* in the sense that they depict mountain landscapes that explode the composition conventions of traditional landscape painting. In Roland's work, one sees fields of rock as rubble or *Geschiebe*.[150]

Restoration and Revival

This structure of repetition and difference – in which the past is input to produce a new future as output – resembles models of genealogy and tradition. It is not aimed at reestablishing some sort of stasis, but rather at managing or controlling change. Despite the novel's focus on "worlds of stone," it does not promote, as some critics have suggested, an aesthetics of "petrification."[151] If anything, the novel takes what is stone or what has been petrified as a starting point for a project of restoration (*Wiederherstellung*) that aims at revival (*Wiederbelebung*). Revival can mean something as sober as reintroducing antiquities into the home (or *Wohnlichkeit*, the milieu of living) or as ecstatic as the aesthetic experience

---

150 On the aesthetics of Stifter's landscapes and landscape painting in relation to natural science, see Karl Möseneder, "Stimmung und Erdleben. Adalbert Stifters Ikonologie der Landschaftsmalerei," in *Adalbert Stifter: Dichter und Maler, Denkmalpfleger und Schulmann: neue Zugänge zu seinem Werk*, ed. Hartmut Laufhütte and Karl Möseneder (Tübingen: Niemeyer, 1996), 18–57, and my article, "Atmosphärische Stimmungen: Landschaft und Meteorologie bei Carus, Goethe und Stifter," *Zeitschrift für Germanistik* 24, no. 2 (2014): 282–95.

151 Cf. Bernd Stiegler, *Philologie des Auges: Die photographische Entdeckung der Welt im 19. Jahrhundert* (Munich: Fink, 2001), 350–364.

Drendorf has during the thunderstorm. Indeed, revival also has a literal sense in the novel, as it refers to the life of the family or to the cultivation of the garden.

As Michael Gamper has suggested, one can read the entire plot of *Indian Summer* as a successful act of calculated matchmaking, through which Risach finds a suitable husband for his surrogate daughter Natalie and thereby repairs his own ill-fated love to Mathilde, Natalie's mother.[152] Risach's garden offers the symbolic evidence for such a reading. At the heart of the garden is an arbor or *Gartenlaube*, a *locus amoenus* that symbolizes traditional notions of paradise, as much as it does, via the rose, the scenery of the *Roman de la rose* and its tale of love and love lost. The rose is also a symbol of Risach's past, a souvenir of his love for Mathilde. In this regard, he symbolically restores this love by recreating the rose arbor, which transfers the past not only into the present, but also into the extended present of a cyclical return. Structured around Drendorf's seasonal visits to the Rose House, the novel also recounts the more regular annual visits of Mathilde with her children to view the roses in bloom. Like the other restorations in the novel, moreover, the rose garden requires countless technical transports and transfers: shipping of seeds, transplants of shoots, hybridization, collection of soil and clippings from various locations, etc. Finally, the roses are most impressive as an arrangement, a wall whose construction requires that the roses be "versezt, verwechselt, beschnitten und dergleichen, bis sich die Wand allgemach erfüllte" [transplanted, changed, pruned, and [...] other things [...] until the wall gradually filled out].[153] If one can read, following Thomas Keller, the wall of roses as a symbol of writing, then the poetological meaning of "restoration" in Stifter's work comes full circle. Restoration has always already been a process of writing, just as the writing of a historical novel has always been a process of restoration.[154]

As artificial as this writing wall may be, however, Risach need not fear the roses will suffer the same fate as other restored objects, which, as Haag suggests, is to lose their attachment to the life given by their original environment: "den Gegenstand seines Lebens zu berauben, weil er dadurch aus jeder Umgebung gerückt würde, keinen Plaz seines Daseins und also überhaupt kein Dasein hätte" [robbing the object of its existence (*Lebens*) since it would have no

---

152 Michael Gamper, "Meteorologie/Wetter," in *Stifter-Handbuch*, 254–255, and Gamper, *Vorsicht. Emergenz eines Dispositivs der Moderne* (Berlin: Alpheus, 2015).

153 HKG 4.1, 143–144 [86].

154 Thomas Keller, *Die Schrift in Stifters "Nachsommer:" Buchstäblichkeit und Bildlichkeit des Romantextes* (Cologne: Böhlau, 1982), 275–277.

true-to-life setting and hence would have no real existence (*Dasein*)].[155] When it comes to the "life" of roses, the proof is in the blooming, so to speak. What matters is not where the rose is from, but whether it thrives in the place where it is (trans-)planted. In this regard, the symbol for the novel's aesthetic ideal is not simply the arbor or the garden, but indeed the greenhouse, a technical improvement that is the precondition of cultivating exotic plants like the *Cereus Peruvianus*, a cactus that Risach's gardener restores to health and even brings to bloom in time for Heinrich and Natalie's marriage.[156] Indeed, Stifter speaks of the cultivation of cacti in the same context as his conservation work:

> Mein Amt als Schulrath als Conservator für Oberösterrich als Vice-Vorstand des Kunstvereines als Referent des Museums dann meine Liebhabereien als Dichter Maler Restaurateur alter Bilder und Geräthe nebst Gerumpel, wozu mich noch im vorigen Sommer die Cactusnarrheit überfallen hat, reiben wahrhaftig eine Riesennatur auf, um so viel mehr als die meinige.
>
> My duties as school councilor as conservator for Upper Austria as vice president of the art association as co-curator of the museum and then my hobbies (*Liebhabereien*) as poet painter restorer of old paintings and furniture including junk, in addition to which a cactus obsession overcame me last summer, would truly drain the energies a giant, so much greater than my own.[157]

This "cactus obsession," however, has little to do with past loss. On the contrary, as Christian van der Steeg has shown, the cactus demonstrates Stifter's interest in the overseas expeditions of his day, a key step in expanding the European imagination into a globalist vision.[158] Earth history not only expands the imagination into the deep past, but also encompasses the earth in its entirety.[159] In the end, the figure who represents this expanding horizon best is Drendorf, a char-

155 HKG 4.2, 242 [309]. Haag, "Versetzt. Restaurierung als Entortung," 85, quotes this passage as the conclusion to her essay. In the passage, Drendorf speaks about his painting of the wood carvings and the regrettable fact that he cannot paint them "from life" in their original setting.
156 Risach is here indeed a "*homo gubernans*," a gardener and custodian of the earth, as argued by Alexander Phillips, "Adalbert Stifter's Alternative Anthropocene: Reimagining Social Nature in 'Brigitta' and 'Abdias,'" in *German Ecocriticism in the Anthropocene*, ed. Heather I. Sullivan and Caroline Schaumann (New York: Palgrave Macmillan, 2017), 66. Zumbusch notes how the "retardation" of the cactus's blooming corresponds with a narrative recalibration of time, "Nachgetragene Ursprünge," 295. Risach experiences love both too early and too late; for Drendorf, everything happens just in time.
157 PRA 18, 270 f.
158 Van der Steeg, *Wissenskunst*, 58–79.
159 In this sense, *Indian Summer* represents a further step in the "geographic turn" described by Tang, *The Geographic Imagination of Modernity.*

acter who has never mourned for a lost rose, but who nonetheless finds a blooming cactus.

Chapter 3
History of a Restorator: Eugène-Emmanuel Viollet-le-Duc

Viollet-le-Duc's Geological Imagination

In Adalbert Stifter's novel *Indian Summer*, the historical conservation movement begins with a small group of antiquarians in the idyllic foothills of Upper Austria. In France, a similar movement arose in the literary and artistic salons of bourgeois Paris. Even as a child, the future architect and defining figure of architectural restoration in France, Eugène-Emmanuel Viollet-le-Duc, found himself at the center of this new development. The first two men to hold the official post of *Inspector General of Historical Monuments*, created in 1830, Ludovic Vitet and Prosper Mérimée, both frequented the salons held by Viollet-le-Duc's father Emmanuel, a writer, and his uncle, Étienne-Jean Delécluze, a painter, at their large family residence on the rue Chabanais.[1] The mentorship of Mérimée, in particular, would help the young Viollet-le-Duc to his first commissions as a restorator of the Romanesque basilica in Vézeley, in 1840, as well as of the Sainte-Chapelle, in collaboration with Félix Duban and Jean-Baptiste-Antoine Lassus, who would also work with Viollet-le-Duc on the restoration of the Notre Dame cathedral in Paris.

Prominent symbols of French religious and national identity, such monuments count among the most important of the French Middle Ages. Viollet-le-Duc had thus established himself, before his thirtieth birthday, as the preeminent conservation architect in France. He solidified this reputation with tireless work on premier projects across the country, from the royal basilica at Saint Denis, to cathedrals in Amiens, Clermont, Munster, Strasbourg, Toulouse, among many others, as well as secular monuments such as the castle at Pierrefonds, the fortified city of Carcassonne, and the Hôtel de Ville in Compiègne.

One of the best measures of Viollet-le-Duc's success and influence is the virulence of the criticism leveled against him. After his death, Viollet-le-Duc's name became synonymous with historical naïveté and restorative excess. In over three hundred pages published in the *Bulletin Monumental* in 1880 and 1881, Anthyme Saint-Paul devastated Viollet-le-Duc's claims to historically authentic restora-

1 The most comprehensive biography to date, and the source of biographical details throughout this chapter, including this information on the salons at the rue Chabanais, is Martin Bressani, *Architecture and the Historical Imagination*, 15.

https://doi.org/10.1515/9783110685534-005

tions by systematically attacking the architect's research methods and subjective attachment to the Gothic style.[2] Later, it became commonplace to dismiss the restorator's idealism as a fantasy, his interventions as vandalism. Anatole France, for example, attacked Viollet-le-Duc's restoration of the castle at Pierrefonds: "Et l'on a détruit des ruines, ce qui est une manière de vandalisme" [And one destroyed the ruins, which is a kind of vandalism].[3] In its completeness, the historical castle appears new, resembling an "énorme joujou" [an enormous plaything].[4] Viollet-le-Duc's restorations, in such a view, are Disneyfications *avant la lettre*. Indeed, like Disney, his name became commonplace, or as Louis Grodecki writes in the *Encyclopedia Universalis* in 1973: "dans l'opinion courante, son nom est associé, en France tout au moins, aux excès romantiques du complément des édifices anciens: 'faire du Viollet-le-Duc' signifie 'restituer abusivement'" [in public opinion, at least in France, his name is associated with romantic excesses in completing antique buildings: 'to do Viollet-le-Duc' means 'to restore to the point of abuse'].[5]

Although the cliché of Viollet-le-Duc's name holds to some extent in public opinion even today, scholars have since reevaluated the architect's contributions to the preservation of French cultural heritage. By the time the Grand Palais hosted an exhibition on Viollet-le-Duc in 1980, Pierre-Marie Auzas, who felt quite alone in publishing a study of Viollet-le-Duc in 1965, could express with confidence: "Viollet le Duc n'est plus méconnu. Viollet le Duc n'est plus méprisé" [Viollet-le-Duc is misrecognized no longer. Viollet-le-Duc is despised no longer].[6] Since then, exhibitions and colloquia devoted to the study of Viollet-le-Duc have become a regular occurrence.[7] The rehabilitation of the architect's legacy does

2 Cf. Robin Middleton, "Viollet-Le-Duc in the Twentieth Century – a French and Anglo-Saxon Survey," in *Eugène Emmanuel Viollet-Le-Duc. Internationales Kolloquium Stiftung Bibliothek Werner Oechslin. Einsiedeln* (Zürich: gta Verlag, 2010), 228–251. As Middleton emphasizes, the fame and criticism of Viollet-le-Duc's oeuvre were both more moderate in England and America. Still, his own dissertation, "Viollet-Le-Duc and the Rational Gothic Tradition" (Dissertation, Cambridge, 1958), played no small part in initiating a scholarly reappraisal of the architect in the late twentieth century.

3 Anatole France, "Pierre Nozière," in *Oeuvres*, ed. Marie-Claire Bancquart, vol. 3 (Paris: Gallimard, 1991), 570.

4 Ibid. 568.

5 Cited in Middleton, "Viollet-Le-Duc in the Twentieth Century," 228.

6 Pierre-Marie Auzas, "Préface," in *Actes du colloque international Viollet-le-Duc Paris 1980*, ed. Pierre-Marie Auzas (Paris: Nouvelles Editions Latines, 1982), 8; Auzas, *Eugène Viollet le Duc, 1814-1879* (Paris: Caisse Nationale des Monuments Historiques, 1965).

7 Notably: an exhibit in 1993 with a focus on Viollet-le-Duc's mountain landscapes: Pierre A. Frey and Lisa Grenier, eds., *Viollet-le-Duc et la montagne* (Grenoble: Grénat, 1993), and more re-

not imply an endorsement of his theory or practice of restoration, however. As Jean Michel Leniaud affirms in his study *Viollet-le-Duc ou les délires du système*, no contemporary conservator or restoration architect would admit adherence to Viollet-le-Duc's most famous theorem of the "unity of style."[8]

Although it at times borders on "delirium," Viollet-le-Duc's vivid imagination has drawn scholars to his work.[9] Titled *Architecture and the Historical Imagination*, Martin Bressani's monumental 2014 biography situates Viollet-le-Duc within the context of nineteenth-century historicist culture in France, which he describes, citing Stephan Bann, as the product of a "desire for history."[10] As a biographer, Bressani takes a psychoanalytic approach to understand both Viollet-le-Duc's personal investment in historical conservation and restoration, and the response of the French nation to the destruction of the Revolution and an industrial modernity defined by accelerating change. In particular, the culture of official mourning during the Bourbon Restoration and the early death of Viollet-le-Duc's mother motivates Bresanni's "central thesis [...] that Viollet-le-Duc's life's work must be conceived as the progressive deepening of an identification with the past predicated upon loss, a self-consuming restorative activity that can never be stabilized in any firm object."[11] Intended as a biographical explanation of Viollet-le-Duc's motivations, Bressani's appeal to psychoanalysis, and in particular, Nicolas Abraham and Maria Torok's theory of melancholia, separates Viollet-le-Duc's imagination from the "firm object" to which it presumably refers. While the objects that attracted the attention of the architect-restorator were usually historical – most often medieval or gothic – Bressani's approach demonstrates how Viollet-le-Duc constructed his own image of the past through the very practices of description, drawing, and building that constitute their object by restoring it.

---

cently a major exhibit at the *Cité de l'architecture et du patrimoine* in Paris: Laurence de Finance and Jean-Michel Leniaud, eds., *Viollet-le-Duc: les visions d'un architecte* (Paris: Norma, 2014). The Stiftung Bibliothek Werner Oechslin held a large colloquium in 2001; Françoise Bercé, *Viollet-le-Duc* (Paris: Éditions du patrimoine, 2013) is a richly illustrated monograph written in the spirit of a museum catalogue.

8 Jean-Michel Leniaud, *Viollet-le-Duc, ou, Les délires du système* (Paris: Mengès, 1994), 5. As the title suggests, Leniaud's biographical study sketches a more negative portrait of its subject than other research in the 1990s. For a critical view of Leniaud, cf. François Loyer, "Préface," in Baridon, *L'imaginaire scientifique*, 9.

9 See, for an early example of scholarly attention to Viollet-le-Duc's imagination, Françoise Boudon, "Le réel et l'imaginaire chez Viollet-le-Duc: les figures du 'Dictionaire de l'architecture,'" *Revue de l'Art* 58 – 59 (1983): 95 – 114.

10 Bressani, *Architecture and the Historical Imagination*, xxiv.

11 Ibid. 27.

In an earlier study, Laurent Baridon analyzes Viollet-le-Duc's "scientific imaginary."[12] Despite his rationalist theory of architecture, Baridon argues, Viollet-le-Duc is at his core an artist. For Baridon, the imaginary is above all a creative space of unification and synthesis. Viollet-le-Duc's practice of drawing establishes imagination in the sense of *image-making* as his preferred method of theoretical analysis. Baridon also compiles an encyclopedia of key concepts and disciplines of scientific knowledge that Viollet-le-Duc drew upon in elaborating a theory and practice of restoration. In addition to terms related to the practice of writing and drawing, the entries include anatomy, botany, crystallography, evolutionism, fauna, geology, milieu, mountain, and organicism – a selection that confirms the predominance of natural history and geology in the architect's imaginary, established by previous research.[13]

Building on this approach, I discuss in this chapter Viollet-le-Duc's geological or geohistorical imagination. I thus understand "geology" not as a subtopic in Viollet-le-Duc's scientific imagination so much as the name for the overarching discursive context of his historical approach. Viollet-le-Duc uses the name "Cuvier," referring to the famous paleontologist and geologist Georges Cuvier (see introduction), as shorthand for this context.[14] In the article "Restoration" in his multi-volume *Dictionary of French Architecture from the Eleventh to the Sixteenth Century*, Viollet-le-Duc traces the genealogy of his own practice of restoration back to Cuvier, praising him as follows:

> Cuvier, par ses travaux sur l'anatomie comparée, par ses recherches géologiques, dévoile tout à coup aux yeux des contemporains l'histoire du monde avant le règne de l'homme. Les imaginations le suivent avec ardeur dans cette nouvelle voie.

12 Laurent Baridon, *L'imaginaire scientifique.*

13 Cf. Bernard Thaon, "Viollet le Duc, pensée scientifique et pensée architecturale," in *Actes du colloque international Viollet-le-Duc Paris 1980*, ed. Pierre-Marie Auzas (Paris: Nouvelles Editions Latines, 1982), 131–142; Françoise Boudon, "Le réel et l'imaginaire chez Viollet-le-Duc," 95–114; and Martin Bressani, "Science, histoire et archéologie. sources et généalogie de la pensée organiciste de Viollet-Le-Duc" (Université de Paris-Sorbonne, 1997), which Baridon cites in its earlier form as a master's thesis. More recently, cf. Jean Nayrolles, "Sciences naturelles et archéologie médiévale au XIXe siècle," in *L'architecture, les sciences et la culture de l'histoire au XIXe siècle*, ed. Centre d'études foréziennes and Ecole nationale supérieure d'architecture (Saint-Etienne) (Saint-Etienne: P.U. de Saint-Etienne, 2001), 25–49.

14 For a general discussion of the relationship between biological and geological thought in the nineteenth century, see Caroline van Eck, *Organicism in Nineteenth-Century Architecture: An Inquiry into Its Theoretical and Philosophical Background* (Amsterdam: Architectura & Natura Press, 1994).

> Cuvier, by his works on comparative anatomy, and by his geological researches, unveiled all at once to the eyes of his contemporaries the history of the world before the reign of man. Imagination follows him with eagerness along this novel path.[15]

The order in which Viollet-le-Duc refers to Cuvier's achievements is significant. Comparative anatomy is shorthand for Cuvier's paleontological practice of anatomical reconstruction that Viollet-le-Duc adapts as a model for architectural reconstruction. Cuvier's geology, however, is both more fundamental and more unsettling than his method of comparative anatomy. Geology is fundamental because it establishes that the world *has* a history "before the reign of man." At the same time, the phenomenon of constant geological change unsettles all notions of a stable present. Cuvier, in particular, is associated with the geological theory of catastrophism.[16] Writing in the wake of the Revolution, France's premier geologist imagined earth history as a series of catastrophic geological revolutions. By aligning himself with Cuvier, Viollet-le-Duc thus not only seeks to acquire scientific legitimacy, but also ascribes, at least implicitly, to a catastrophic theory of history that motivates the practice of restoration.

In the everyday work of the historical conservationist, moreover, geological forces from weather to earthquakes are an implacable enemy. While critics admire Viollet-le-Duc for his successful structural repairs to historical monuments, which began with his very first independent restoration project in Vézelay, his consciousness of the geological inevitability of structural collapse has received less attention. This broad conceptual grounding of conservation work in geological science is the subject of this chapter.

This chapter thus begins with a close reading of Viollet-le-Duc's programmatic article "Restoration," which leads to an examination of how the architect adapts Cuvier's earth-historical research to the history of architecture. Second, expanding upon Bressani's biographical approach, I argue that the loss that motivates Viollet-le-Duc's restorative activities is not only personal and national, but also geological. In context of his lifelong fascination with geology, I discuss thirdly how the practice of drawing is central to all aspects of Viollet-le-Duc's imagination. Indeed, for the architect, imagination is not so much a psychological faculty as an artistic practice: the category of the imaginary encompasses all that can be drawn. Viollet-le-Duc's geological imagination provides, finally, the basis for a comparison with Adalbert Stifter's literary representation of restoration practice in *Indian Summer.* This comparison also provides a transition to

15 Viollet-le-Duc, "Restauration," 15 [13–14].
16 Even if historians of science now consider Cuvier's work on fossils the more important achievement, cf. Martin Rudwick, *Georges Cuvier, Fossil Bones, and Geological Catastrophes*, 266.

this study's final chapter, which focuses on the relationship between geohistoricism and the reconstruction of history in Stifter's historical novel *Witiko.*

Defining Restoration: Viollet-le-Duc's *Dictionary of French Architecture*

Viollet-le-Duc's definition of "Restoration" appeared in the eighth volume of his monumental *Dictionnaire raisonné de l'architecture française du XIe au XVIe siècle* in 1866 – eight years after the appearance of the first volume, and two before the completion of the tenth and final volume of the first edition in 1868. Of his theoretical writings, the *Dictionary of French Architecture* is undoubtedly Viollet-le-Duc's magnum opus. All told, its 427 articles span over 5,000 pages, including 3,745 figures.[17] With this vast number of illustrations, the *Dictionary* is primarily a visual work. It also has a pragmatic emphasis: more a handbook of medieval architecture than a history. Written during a successful period of Viollet-le-Duc's career, the dictionary thus illustrates and promulgates the principles of his architectural practice. As a friend of Napoleon III, Viollet-le-Duc not only thrived professionally in the Second Empire, he also found in the Second Empire's broader political project of "restoring" the French nation a new legitimacy and energy, as well as material support for his restoration projects. As Bressani notes, the first announcement for the *Dictionary* came just weeks after Viollet-le-Duc was named, on March 7, 1853, *Inspecteur général des édifices diocésains* [Inspector General of Diocesan Buildings], a decisive consolidation of his powerful position within the French patrimonial institutions.

As a practical work, the vast majority of the articles address individual elements of medieval architecture. One typical series from the first volume runs: *Accolade, Accoudoir, Agrafe, Aiguille, Albâtre* (respectively: a curved ornament above doors and windows, an armrest, a metal pin, the "needle"-form of clocktowers, alabaster). Still, a small number of articles address broader theoretical issues. The headings "Architecture" and "Construction" (each with various subheadings) introduce treatises of more than three hundred pages each.[18] Structured, like all of the articles, by redundancy and internal cross-reference,

17 Bressani, *Architecture and the Historical Imagination*, 231. On the publishing history of the work, 228–232.

18 On the constitutive function of the distinction between "architecture" and "construction" in Viollet-le-Duc's system, cf. Hubert Damisch, "Introduction," in *Viollet-le-Duc, L'architecture raisonnée, extraits du "Dictionnaire raisonné de l'architecture,"* by Eugène-Emmanuel Viollet-le-Duc, ed. Hubert Damisch (Paris: Hermann, 1964), 9–26.

these articles represent Viollet-le-Duc's attempt to provide a coherent and universal theory of medieval French architecture.

Among the theoretical and programmatic texts in the *Dictionary*, the twenty-page article on "Restoration" is thus comparatively short. Published in 1866, it also appears later than the systematic treatises on "Architecture" and "Construction." If the *Dictionary* as a whole can be read as a statement of architectural theory written at the high point of Viollet-le-Duc's career, the article "Restoration" can be read as a retrospective reflection on his life's work as a restorator, within the context of the national project of historical conservation. Accordingly, the article recounts the history of restoration in France as a story of progress and success. After quoting at length Ludovic Vitet's report on the conservation of historical monuments from 1831, a foundational document by France's first Inspector General of Historical Monuments, Viollet-le-Duc concludes, with a reference to his own restoration of the castle of Pierrefonds:

> Ce programme si vivement tracé par l'illustre critique il y a trente-quatre ans, nous le voyons réalisé aujourd'hui, non sur le papier, non par des dessins fugitifs, mais en pierre, en bois et en fer pour un château non moins intéressant, celui de Pierrefonds.
>
> The program so vigorously sketched by the illustrious critic thirty-four years ago, we now see realized, not on paper, but in stone, wood and iron in a castle not less interesting, that of Pierrefonds.[19]

Viollet-le-Duc's article "Restoration" thus legitimizes his own restoration work, at Pierrefonds and elsewhere, retrospectively as the fulfillment of a national project, institutionalized as the Historical Monuments Commission.

This context helps explain Viollet-le-Duc's famous and yet still surprising claim that the goal of architectural restoration is "un état complet qui peut n'avoir jamais existé à un moment donné" [a condition of completeness that may never have existed at any given time].[20] While Viollet-le-Duc appears to admit here that restorations are inauthentic and ahistorical, the notion of "completeness" is inseparable from the uniqueness of his own historical moment. Indeed, only the nineteenth century has the historical knowledge required for restoration, as Viollet-le-Duc understands it. At the start of his article, Viollet-le-Duc establishes the modernity of the concept: "Le mot et la chose sont mod-

---

19 Viollet-le-Duc, "Restauration," 20 [27]; In his report, Vitet had envisioned restoring the castle of Coucy.

20 Ibid. 14 [9]. Translation modified from "could never have existed" since Viollet-le-Duc's phrase only suggests that the historical existence of a "complete condition" is uncertain.

ernes" [The term Restoration and the thing itself are both modern].[21] As mentioned above, he dates the advent of the concept to the second quarter of the nineteenth century, and more specifically, to the naming of Vitet as Inspector General of Historical Monuments in 1830. In Viollet-le-Duc's view, the advent of restoration in France is a singular event in world history: "aucune civilisation, aucun people, dans les temps écoulés, n'a entendu faire des restaurations comme nous les comprenons aujourd'hui" [no civilization, no people of bygone ages, has conceived the idea of making restorations in the sense which we comprehend them].[22] He supports this view with ethnographic as well as philological arguments. In Asia, for example, instead of restoring an old temple, one lets it be and constructs a new one beside it. The Romans would not have recognized the word "restauration," Violet-le-Duc claims, only "restitution" (*restituere*), and words such as "instaurare, reficere, renovare" which "ne veulent pas dire restaurer, mais rétablir, refaire à neuf" [do not mean to restore, but to reinstate – to make anew].[23]

Important here is not so much the validity of the architect's arguments as the claim to historical uniqueness.[24] For the uniquely modern practice also produces unique results. Here is Viollet-le-Duc's full definition:

> Restaurer un édifice, ce n'est pas l'entretenir, le réparer ou le refaire, c'est le rétablir dans un état complet qui peut n'avoir jamais existé à un moment donné.
>
> To restore a building is not to preserve it, to repair, or rebuild it; it is to reinstate it in a condition of completeness which may never have existed at any given time.[25]

This definition by negation suggests that one understands restoration best by exclusion. *Entretenir* or "preserve" suggests that a building never really suffers degradation, at least not in its essence, while *réparer* or "repair" implies that any damage is ultimately reversible. The proximity of restoration to rebuilding (*refaire*) is a particularly sensitive point. Viollet-le-Duc demonstrates his commitment to the value of history by rejecting the idea of rebuilding (and, as he argues later in the essay, of the neogothic style in modern buildings). Yet Viollet-le-Duc also holds onto an idealistic belief in the "complete condition" of the monument. This belief holds the aesthetic value of art higher than the fact of historical

21 Ibid.
22 Ibid. 10 [14].
23 Ibid.
24 In fact, the Romans did use the word *restauro* and *restauratio*, though *instauro* is more common and older, cf. Charlton T. Lewis and Charles Short, "Restauro" and "Restauratio," in *A Latin Dictionary* (Oxford: Clarendon, 1879).
25 Viollet-le-Duc, "Restauration," 14 [9].

change. As Viollet-le-Duc knows, most medieval cathedrals and castles were built over decades or even centuries and in most cases were never completed. The "complete condition" is a heuristic fiction.

By embracing positively this idealist position, Viollet-le-Duc thus addresses head on the his critic's primary contention against him. For example, his contemporary John Ruskin was strongly opposed to the practice of restoration:

> Neither by the public, nor by those who have the care of public monuments, is the true meaning of the word *restoration* understood. It means the most total destruction which a building can suffer: a destruction out of which no remnants can be gathered: a destruction accompanied with false description of the thing destroyed. Do not let us deceive ourselves in this important matter; it is impossible, as impossible as to raise the dead, to restore anything that has ever been great or beautiful in architecture.[26]

Ruskin's "total destruction" and Viollet-le-Duc's "complete condition" are two sides of the same coin. Ruskin's critique is so devastating because it identifies the central paradox of restoration: in the act of saving a historical monument, restoration destroys the aura of authenticity that made it valuable in the first place. By contrast, Viollet-le-Duc's definition transforms the impossibility of restoration into an ideal. It transforms irretrievable loss into a virtual state of perfection, which may never have existed and may or may not be achievable.

Following Cuvier: From Archeology to Reconstructive Surgery

Despite this idealism, Viollet-le-Duc would not have accepted the criticism that restoration destroys the "remnants" of a building. On the contrary, "remnants" are the basis of every restoration, which Viollet-le-Duc presents as a synthesis between architecture and archeology. He thus advises future restorators:

> avant de commencer, faut-il tout fouiller, tout examiner, réunir les moindres fragments en ayant le soin de constater le point où ils ont été découverts, et ne se mettre à l'œuvre que quand tous ces débris ont trouvé logiquement leur destination et leur place, comme les morceaux d'un jeu de patience.

---

26 John Ruskin, *The Works of John Ruskin*, ed. E.T. Cook and A. Wedderburn, vol. 8, 39 vols. (London: George Allen, 1903), 242. Ruskin does not address any specific restorators or restorations, though his main target is likely Gilbert Scott, cf. Norbert Huse, *Denkmalpflege: Deutsche Texte aus drei Jahrhunderten* (Munich: C.H. Beck, 1984), 90. While focusing on the German context, Huse identifies Ruskin as a key voice in the emerging debate between "restoration" and "conservation," ibid., 84–93. By 1900, Georg Dehio had consolidated this debate into a motto, "conserve, do not restore," cf. Dehio and Riegl, *Konservieren, nicht restaurieren.*

> before beginning [a restoration] it will be necessary to search for and examine all that remains; to collect the smallest fragments – taking care to note the point where they were found; and not to begin the work until their place and use have been assigned to all these remains, as with the pieces of a puzzle.[27]

The article "Restoration" thus reconnects with earlier concerns of the Historical Monuments Commission, whose members often emphasized the importance of "conservation" over imaginative restoration. Archeologist Charles Lenormant, for example, expressed in the Commission meetings a sentiment similar to Ruskin's later critique: "the imagination in matters of restoration can only lead to disastrous results."[28] Indeed, even while praising Ludovic Vitet's ground-breaking work at the start of "Restoration," Viollet-le-Duc adopts an attitude of superiority with regards to his predecessor, who was, by comparison, lacking in archeological knowledge.[29]

Although the goal of "restoration" may be a virtual or non-real condition of completeness, Viollet-le-Duc claims that he has grounded his practice of restoration in scientific knowledge. And he appeals not to the authority of archeologists, who themselves only began developing a rigorous scientific methodology in the second half of the nineteenth century, but to the most famous French scientist of his day, Georges Cuvier. Importantly, Viollet-le-Duc credits Cuvier with a radical change in "attitude" towards the past: "Notre temps, et notre temps seulement depuis le commencement des siècles historiques, a pris en face du passé une attitude inusitée" [Our age has adopted an attitude towards the past in which it stands quite alone among historical ages].[30] The architect attributes this new mentality to the sudden shock of deep time, as if the geologist had presented his findings on earth history in a spectacular unveiling ceremony. The reconstructions of geology are so spectacular that "les imaginations le suivent avec ardeur dans cette nouvelle voie" [imagination follows him with eagerness along this novel path].[31] Cuvier not only captures the imagination of his contemporaries, he channels it into a path that proves productive for numerous disciplines:

> Des philologues, après lui, découvrent les origines des langues européennes, toutes sorties d'une même source. Les ethnologues poussent leurs travaux vers l'étude des races et de leurs aptitudes. Puis enfin viennent les archéologues, qui, depuis l'Inde jusqu'à l'Égypte

27 Viollet-le-Duc, "Restauration," 34 [70].
28 Quoted in Bressani, *Architecture and the Historical Imagination*, 105.
29 Viollet-le-Duc, "Restauration," 18 [22].
30 Ibid. 15 [13].
31 Ibid.

et l'Europe, comparent, discutent, séparent les productions d'art, démasquent leurs origines, leurs filiations, et arrivent peu à peu, par la méthode analytique, à les coordonner suivant certaines lois.

Next comes the philologist, who discovers the origin of European languages, all issuing from the same source. The ethnologist extends his labours to the study of races and their aptitudes. Lastly, comes the archeologist, who investigating the productions of art from India to Egypt and Europe, compares, discusses, and discriminates them, unmasking their origins and their affiliations; and by the analytical method succeeds gradually in co-ordinating them according to certain laws.[32]

Each of these three fields plays a specific role in Viollet-le-Duc's historical and scientific imagination. From a historical standpoint, philology picks up where geology leaves off, with the advent of humans as speaking beings; ethnology, in turn, deals with the historical differentiation of humans into races; the archeologist, finally, analyzes the "productions of art" of these different peoples. From a scientific standpoint, each field represents a different adaptation of Cuvier's earth-historical methods for the purposes of human historiography. Archeology represents the global application of the scientific methods of geology to cultural artefacts. The ethnological theory of races relates to Viollet-le-Duc's conception of art as an instinct and of architecture as habitat.[33] In philology, Viollet-le-Duc finds a model for the evolution of culture, the "principle of its changes."[34]

It is not easy to summarize Cuvier's vast and varied impact on the nineteenth-century imagination of (earth) history. As mentioned above, Cuvier did not originate, but certainly popularized the theory of geological catastrophism. His work as a paleontologist was more influential. With its emphasis on functional organization, his comparative anatomy revolutionized biological study, indeed founded biology in the modern sense, as the study of life and not just living beings. In addition to subordinating taxonomy to the functionality of organs, Cuvier's comparative anatomy postulates a principle of correlation for organic forms: the form of each organ determines the form of all other organs belonging to an

---

32 Ibid. 15 [14].

33 In the 1860s, Viollet-le-Duc became increasingly interested in theories of race, including those expressed in the notorious *Essai sur l'inégalité des races humaines* (1853–1855) by Joseph-Arthur de Gobineau. Especially in the *Entretiens sur l'architecture*, next to the two dictionaries the most important work in Viollet-le-Duc's theoretical oeuvre, the architect seeks to explain historical-geographical differences in architecture as a product of racial differences. As the article "Restoration" demonstrates, however, racial theory is not the telos, so much as a side-path of Viollet-le-Duc's organicist theory. For a full discussion of race and architecture in the *Entretiens*, see Bressani, *Architecture and the Historical Imagination*, 345–365.

34 Baridon, *L'imaginaire scientifique*, 218.

animal. This principle provides the basis for Viollet-le-Duc's organicist theory of architecture – it influenced other architects of the period as well, including Gottfried Semper.[35] Finally, Cuvier related the function of organs to the survival of the animal within its milieu, that is, according to its conditions of existence.[36]

Biology and geology intersect in Cuvier's study of fossils, which play a central epistemic role in his work. The fossil is an object that sets in motion various procedures of reconstruction. By definition, it has undergone a process of change (fossilization) that an imaginative act of recognition must first reverse. To understand the epistemic obstacles involved in this basic reconstruction, one need only think of earlier theories that saw fossils as spontaneous formations in living rock.[37] Secondly, as Cuvier's comparative anatomy demonstrates, one can understand fossils or fossilized bones as fragmentary parts of a coherent skeletal whole. These are the reconstructions – the mammoths, dinosaurs, and other prehistoric creatures – that captivated the imagination of Viollet-le-Duc's contemporaries. Thirdly, because Cuvier's anatomy relates organic function to the *milieu*, a skeleton is not merely a puzzle of reassembly, but also a window to the world in which the animal once lived. Paleontology thus also produces a spectacle of prehistoric "scenes," drawings and paintings of worlds previously unimagined.[38] These scenes, in turn, relate back to the geological strata from which they emerged, which the geologist can now conceptualize, and date, as a constellation or collection of fossils.[39]

The fact that Cuvier's analysis of fossils could lead to imaginative reconstruction of prehistoric worlds astonished his contemporaries. In his paean to Cuvier as "le plus grand poète de notre siècle" [the greatest poet of our age], Balzac draws out the contrast between the naturalist's inauspicious material of study and the remarkable results of his research:

> notre immortel naturaliste a reconstruit des mondes avec des os blanchis, a rebâti comme Cadmus des cités avec des dents, a repeuplé mille forêts de tous les mystères de la zoologie avec quelques fragments de houille, a retrouvé des populations de géants dans le pied d'un

35 Cf. Baridon, *L'imaginaire scientifique*, 214; Caroline van Eck, *Organicism*, 231–234.

36 Martin Bressani has associated the relation of the organism to its milieu to Viollet-le-Duc's conception of "architecture," while associating the principle of correlation among parts to the architect's concept of "construction," cf. Bressani, "Opposition et Équilibre: le rationalisme organique de Viollet-Le-Duc," *Revue de l'Art* 112 (Spring 1996): 28–37.

37 Paolo Rossi, *The Dark Abyss of Time*, 7.

38 Cf. Martin Rudwick, *Scenes from Deep Time: Early Pictorial Representations of the Prehistoric World* (Chicago: U Chicago P, 1992).

39 Rudwick traces the history of this use of fossils to François-Xavier Burtin's "Révolutions générales" (1789), cf. *Bursting the Limits of Time*, 200–203.

mammouth. Ces figures se dressent, grandissent et meublent des régions en harmonie avec leurs statures colossales.

our immortal naturalist has reconstructed whole worlds out of bleached bones. Like Cadmus, he has rebuilt great cities from teeth, repopulated thousands of forests with all the mysteries of zoology from a few pieces of coal, discovered races of giants in the foot of a mammoth. These figures rise up and tower high above us, furnishing whole regions in accordance with their colossal stature.[40]

Richly imaginative minds like those of Balzac or Viollet-le-Duc follow the naturalist so eagerly precisely because his reconstructions appear to make such great leaps. At the same time, the most outlandish reconstructions retain a sense of authority by virtue of their association with Cuvier, the age's most famous and reputable scientist, professor at the *Collège de France*, and member of the Royal Society of London.

Cuvier's fame and the widespread popularity of his paleontological and geological reconstructions thus present the scientific and broader cultural-historical backdrop for Viollet-le-Duc's imaginative adaptation of comparative anatomy. In his article "Restoration," Viollet-le-Duc does not so much elaborate his organicist theory of architecture as presuppose it as a necessary scientific foundation of the practice. He treats each architectural monument as a living being, "Nous ne saurions trop le répéter, les monuments du moyen âge sont savamment calculés, leur organisme est délicat" [We cannot repeat it too often: The medieval buildings are planned with deliberate skill; their organism is delicate].[41] This metaphor gives rise to a comparison of the work of restoration to medical surgery:

Dans la structure du moyen âge, toute portion de l'œuvre remplit une fonction et possède une action. C'est à connaître exactement la valeur de l'une et de l'autre que l'architecte doit s'attacher, avant de rien entreprendre. Il doit agir comme l'opérateur adroit et expérimenté, qui ne touche à un organe qu'après avoir acquis une entière connaissance de sa fonction, et qu'après avoir prévu les conséquences immédiates ou futures de son opération. S'il agit au hasard, mieux vaut qu'il s'abstienne. Mieux vaut laisser mourir le malade que le tuer.

In medieval construction every portion of the work fulfills a function and has an action. The architect should make a point of ascertaining the value of both before commencing his undertaking. He should proceed like the skillful and experienced [surgical] operator, who does not touch an organ until he has acquired a thorough acquaintance with its function,

40 Balzac, *La Comédie Humaine*, vol. 10, 75 [19].
41 Viollet-le-Duc, "Restauration," 32 [66].

> and provided for the immediate and remote consequences of his operation. Rather than proceed at hazard he should not undertake it. Rather let the patient die than kill him.[42]

In Viollet-le-Duc's vision, the study of "anatomy" is thus as central to the education of an architect as it is to the future surgeon. To maintain the structural integrity of the building, the architect must understand the specific function of each element in its relation to the functions of other elements, and be able to assess, for example, the counteracting forces of an arc and its pillar.[43] The restorator must know a building intimately, its history, style and form, and "mieux encore, s'il est possible, connaître sa structure, son anatomie, son tempérament, car avant tout il faut qu'il le fasse vivre" [if possible, be still better acquainted with its structure, its anatomy, its temperament; for it is essential above all things that he should make it live].[44] Viollet-le-Duc also likens early attempts at restoration to medical students' rough first tries at anatomical dissection: "Cette malheureuse église de Saint-Denis fut comme le cadavre sur lequel s'exercèrent les premiers artistes entrant dans la voie des restaurations. Pendant trente ans elle subit toutes les mutilations possibles" [This unfortunate church of St. Denis was a sort of anatomical subject on which artists who first entered on the path of restoration made their first essays in restoration. During thirty years it suffered every possible mutilation].[45]

Such analogies to human anatomy lend Viollet-le-Duc's article on restoration rhetorical verve, but they also have an epistemological basis in the anatomical works of Cuvier. Viollet-le-Duc imagines restoration as the direct application of the principles of comparative anatomy and the methods of reconstructive paleontology to architecture:

> L'architecture du moyen âge, au moment où elle abandonne les traditions abâtardies de l'antiquité, c'est-à-dire du XII au XVème siècle, possède le style, parce que, plus qu'une autre peut-être, elle procède avec cet ordre logique que nous entrevoyons dans les œuvres

42 Ibid. 33 [67].
43 To avoid confusion, it is worth emphasizing that the word "function" here refers exclusively to structural functionality of architectural elements such as the arc and its pillar, or the capital with its column and base. Other texts, notably the *Histoires* (see below) employ "function" to mean the cultural use of a building. On the role of "structure" and "function" in Viollet-le-Duc's thought, cf. Hubert Damisch, "Les 'Entretiens sur l'architecture' ou du structuralisme au fonctionnalisme," in *Actes du colloque international Viollet-le-Duc Paris 1980*, ed. Pierre-Marie Auzas (Paris: Nouvelles Editions Latines, 1982), 93–99; and Damisch's earlier, controversial reading of the architect as a proto-structuralist, "Introduction," 9–26.
44 Viollet-le-Duc, "Restauration," 27 [49].
45 Ibid. 22 [34–35].

de la nature. Aussi de même qu'en voyant la feuille d'une plante, on en déduit la plante en entière; l'os d'un animal, l'animal entier: en voyant un profil, on en déduit les membres d'architecture, du membre d'architecture le monument.

Medieval architecture, as soon as it abandons the bastardized traditions of antiquity, that is, from the twelfth to the fifteenth century, possesses style because, perhaps more than any other architecture, it proceeds according to the logical order that we observe in the works of nature. And just as, from looking at the leaf of a plant, one can deduce the entire plant; from an animal bone, the entire animal: looking at a profile, one can deduce the architectural limbs, from the limb, one can deduce the entire monument.[46]

This explanation of the anatomical deduction of architectural forms appears in the article "Style" in the *Dictionary of French Architecture.* In this context, Viollet-le-Duc takes pains to distinguish the concept of "style" in the singular from the more common notion of plural "styles" or "fashions."[47] If Viollet-le-Duc's distaste for the architecture of Greco-Roman antiquity is in keeping with his romantic and nationalist preference for French medievalism, the term "bastardized" (*abâtardies*) highlights his belief that architecture develops through quasi-natural processes. In Viollet-le-Duc's architectural theory, the unity of style in architecture is like a purebred race of dog or a select variety of rose, except that the architect imagines nature itself as the best selector of purity. Style is thus unified and universal when it follows the unique logic of the works of nature. The organic metaphor is so self-evident for Viollet-le-Duc that he does not elaborate its theoretical foundations at any length. As Bressani has investigated in detail, its use was widespread among writers such as Ludovic Vitet, Louis Bâtisser, Daniel Ramée, and Eugène-Joseph Woillez, who were all associated with the Historical Monuments Commission or personally close to Viollet-le-Duc.[48]

The comparison of architectural forms to nature, of course, is much older. In his *Essay of Architecture*, a classic of French architectural theory first published in 1753, for example, Abbé Laugier solidifies the paradigm of architecture imitating nature in the image of the primordial house as a construction modeled on the shape of trees. The book's frontispiece depicts this original house quite literally as an extension of live, growing trees.[49] In *Von deutscher Baukunst* [On German Architecture], Goethe famously describes the vaults in Erwin of Steinbach's Strasbourg cathedral as "Bäume Gottes" [divine trees], and, as Laurent Baridon argues, Sir James Hall's experiments in modeling Gothic forms in a wicker gar-

46 Viollet-le-Duc, "Style," in: *Dictionnaire raisonné de l'architecture*, vol. 8, 486.

47 Cf. Bressani, *Architecture and the Historical Imagination*, 381–382.

48 Bressani, *Architecture and the Historical Imagination*, 270, and his earlier dissertation, "Science, Histoire et Archéologie."

49 Viollet-le-Duc owned a copy of the 1755 edition, cf. Baridon, *L'imaginaire scientifique*, 232.

den arbor completed the circle of the analogy.[50] Although Hall may be better known for his geological experiments, which sought to provide an empirical basis for the geohistorical theories of his friend James Hutton, his experiments with tree and vine constructions, which he explicitly calls a "restoration" of Gothic forms, demonstrate how readily geological insights transferred to archeological and art historical speculations circa 1800.[51]

Hall's experiments are paradigmatic of how art not only imitates nature's outer appearance, as a *natura naturata*, but also its inner structure and the dynamic principles of its development, or *natura naturans*. As Baridon emphasizes, Viollet-le-Duc's organicism seeks above all to explain how Gothic ornamentation is inseparable from its structure. Only then does the talk of architectural "anatomy" acquire its full meaning as the dynamic interrelation of structural forces. The inner rationality of Viollet-le-Duc's comparative anatomy of archeological method thus also depends on the relationship between form and function that underlies Cuvier's biological reconstructions. Functional rationality allows Viollet-le-Duc to consider his method scientific, even as he insists on the priority of ideal forms over existing buildings. The idealism of his conception of architecture is also evident in the 'ideal' Gothic cathedral he sketched for the *Dictionary of French Architecture*, which one can also view as an ideal restoration a building that "may never have before existed." An amalgam of various existing buildings, the sketch of the cathedral also illustrates its ideal in the same way that an anatomical model represents the ideal human body.[52]

Viollet-le-Duc's anatomical theory of architecture is systematic and idealistic, but it also has a historical dimension. In the shift from a synchronic to a diachronic analysis of architectural forms, Viollet-le-Duc develops an evolutionary theory of architecture.[53] To find inspiration for this endeavor, however, he must look beyond the theories of Georges Cuvier. Not surprisingly, the architect

---

50 Cf. Baridon, *L'imaginaire scientifique*, 210. Johann Wolfgang von Goethe, "Von deutscher Baukunst. D.M. Ervini a Steinbach. 1773," in *Sämtliche Werke*, ed. Friedmar Apel, vol. 18 (Frankfurt a. M.: Deutscher Klassiker Verlag, 1998), 110.

51 Sir James Hall, *Essay on the Origin, History, and Principles of Gothic Architecture* (London: W. Bulmer, 1813), 19. The book's frontispiece also depicts the "wicker fabric" that Hall constructed in his garden. See also Joseph Rykwert, *On Adam's House in Paradise: The Idea of the Primitive Hut in Architectural History* (Greenwich, CT: Museum of Modern Art, 1972), 83 f.

52 For a recent take on the "ideal cathedral," see Peter Kurmann, "Viollet-le-Duc und die Vorstellung einer idealen Kathedrale," in *Eugène Emmanuel Viollet-le-Duc. Internationales Kolloquium Stiftung Bibliothek Werner Oechslin*, 32–50.

53 Cf. Laurent Baridon, "Anatomie comparée et pensée évolutionniste dans la théorie et la pratique architecturale de Viollet-le-Duc," in *L'architecture, les sciences et la culture de l'histoire au XIXe siècle*, ed. Loyer, 73–81.

turns instead to theses of Étienne Geoffroy Saint-Hilaire, Cuvier's opponent in the "debate of 1830," a sensational event that represented not only the last stand of Cuvier's fixed models of natural history, but also carried all the political energy of a generational conflict. Geoffroy Saint-Hilaire's theory of the homology of forms, which takes up and extends Goethe's earlier morphological concepts, resonates with Viollet-le-Duc's idealist rationality. In his later *Histoire de l'habitation humaine* (1870), Viollet-le-Duc writes a fuller evolutionary history of architecture, but the *Dictionary of French Architecture* already depends on evolutionary ideas, even if it remains loyal to its systematic format and the accompanying argument of the coherence of medieval architecture. Indeed, Viollet-le-Duc appeals to the coherent unity of this architecture to justify the strict evolutionary logic of its art-historical development:

> Cet art [...] ne procède point, comme de nos jours, par une succession de *modes*, mais par une filiation non interrompue dans l'application des principes admis. Si bien qu'en décomposant un édifice du XVème siècle, on peut y retrouver le développement de ce que ceux du XIIème siècle donnent en germe, et que, en présentant une suite d'exemples choisis entre ces deux époques extrêmes, on ne saurait, en aucun point, marquer une interruption. De même, dans l'ordre de la création, l'anatomie comparée présente, dans la succession des êtres organisés, une échelle dont les degrés sont à peine sensibles, et qui nous conduits, sans soubresauts, du reptile jusqu'à l'homme.
>
> This art does not proceed, as in our day, by a succession of *fashions*, but instead by an uninterrupted line (*filiation*) in the application of the accepted principles. Just as, by analyzing a building from the fifteenth century, one can find the development of what the buildings of the twelfth century produce as a seed (*donnent en germe*), and just as, by presenting a series of examples chosen between these two periods, one cannot mark an interruption at any point, so, in the order of creation, comparative anatomy presents, in the succession of organic beings, a ladder whose steps are hardly visible, and which leads us, without any leaps, from the reptile to the human being.[54]

Once again, Viollet-le-Duc emphasizes the distinction between the "succession of fashions" that appears as the rule of contemporary culture, and the "principles" of medieval art. Again, these principles are strictly analogous to the "filiations" and "germinations" of the natural world.

The presumed existence of a "ladder" with infinitesimally small steps, a series of examples that connects two epochs without "interruption" not only accords with Geoffroy Saint-Hilaire's morphological theories, it also has important practical implications for Viollet-le-Duc's work. Indeed, Viollet-le-Duc's theoret-

54 Viollet-le-Duc, "Profil," in: *Dictionnaire raisonné de l'architecture*, vol. 7 (1864), 521–522; quoted in Baridon, *L'imaginaire scientifique*, 39.

ical framework is inseparable from the practice of restoration. Because there are no jumps in the history of architecture, Viollet-le-Duc can speculatively fill in the missing pieces of the puzzle that a damaged building presents to him. In the article "Restoration," Viollet-le-Duc thus cites Ludovic Vitet's *Monographie de l'église Notre-Dame de Noyon* (1845), which envisions a comparative art history as a diachronic geography of architectural forms. Vitet calls the result of this history a "tableau comparatif" [comparative tableau], with axes for time and locality that trace the changes of each art or architectural form.[55] The practical use of this comprehensive tableau becomes clear in the context of the central dilemma in Viollet-le-Duc's practice of restoration, which he illustrates with an example of a broken "chapiteau" [capital] from a twelfth-century building.[56] The problem is not merely the damage to the capital, but also the repairs undertaken in the intervening centuries. For the nineteenth-century restorator often finds a capital from the thirteenth, fourteenth, or fifteenth century in a twelfth-century cathedral. In this case, Viollet-le-Duc faces a dilemma: he can restore the building as is, as a "*fac-simile*," but this would "produire les plus graves erreurs, en consacrant, pour ainsi dire, une interpolation" [lead to the gravest errors, by consecrating what may be called an interpolation]. In the second case, "la substitution d'une forme première à une forme existante, reconnue postérieure, fait également disparaitre les traces d'une réparation" [the substitution of a primary form for an existing one of a later period also obliterates the traces of a reparation].[57] In either case, however, a comparative tableau is necessary: first, to identify the hybrid origins of architectural elements, and second, to make possible "the substitution of a primary form," if this is the path chosen. Towards the end of the essay, Viollet-le-Duc introduces a further analogy: the architect-restorator must above all know the "grammar" of previous epochs of building. Medieval art "s'appuie sur des principes, et non sur un formulaire; comme une langue bien faite peut exprimer toutes les idées sans faillir à sa grammaire" [rests on principles [...] in the same way as a competent language can express any idea without sacrificing grammatical consistency].[58] Viollet-le-Duc's *Dictionary of French Architecture* thus presents as much the grammar as the lexicon of medieval architectural forms. Despite this insistence that medieval architecture relies on principles, which, once learned, can generate new, presumably authentic con-

55 Viollet-le-Duc, "Restauration," 20 [28], translation modified from "comparative sketch." Like the *scala naturae*, the notion of the tableau suggests, as in natural history or grammar tables, that one can deduce missing forms.

56 Ibid. 15 [12].

57 Ibid.

58 Ibid. 32 [64].

structions, Viollet-le-Duc admits that when it comes to creative acts of restoration, "la pente est glissante" [we are on slippery ground].[59]

Ludovic Vitet's emphasis on the importance of local differences in the development of artistic forms can serve to limit the creative license of the architect who learns the "grammar" of medieval architecture. This grammar is not universal, but rather dispersed across a variety of local dialects. In Viollet-le-Duc's adaptation of biological models, the concept of milieu plays a similar role. For Viollet-le-Duc, a monument is always the product of the "le sol qui l'a vu naitre" [the land where it was born].[60] Art is "culture" in the etymological sense, and paradoxically, this understanding brings us back to the realm of nature: "L'art est un arbre qu'on n'élague pas et qui n'a pas besoin de tuteurs. Il ne pousse qu'en terre libre, en prenant sa sève comme il peut et où il veut, en développant ses rameaux en raison de sa nature propre" [Art is a tree that does not need pruning or stakes for support. It only grows on free land, taking its sap as it can and wherever it wants, developing its branches according to its own nature].[61] In growing "according to its own nature," art must nonetheless adapt, like natural trees, to environmental conditions. In the article "Profile," Viollet-le-Duc explains with exacting precision how the angle of the sun and quantities of rain determine the evolution of the form of a capital. This determinism combines Cuvier's functional analysis of organs with Lamarck's ideas of evolution resulting from environmental adaptation.[62]

Interiors: The Milieu of Everyday Life

If architecture evolves in relation to its outer environment, then buildings may themselves be environments, which influence the development of their interiors. In addition to the ten volumes of the *Dictionary of French Architecture* (1854–1868), Viollet-le-Duc also published in parallel a six-volume *Dictionary of Furniture* (1858–1875). The *Dictionary of Furniture* has a remarkable structure. Following the expected alphabetical treatment of individual furnishing types, a second section demonstrates the historiographical implications of this "interior turn." As monuments represent places of memory for significant public events, furni-

59 Ibid.

60 E.E. Viollet-le-Duc, "Architecture," in: *Dictionnaire raisonné de l'architecture*, vol. 1 (Paris: B. Bance, 1854), 146; quoted in Baridon, *L'imaginaire scientifique*, 191.

61 E.E. Viollet-le-Duc: "Sculpture," in: *Dictionnaire raisonné de l'architecture*, vol. 8, 249–251; quoted in Baridon, *L'imaginaire scientifique*, 273.

62 Cf. Baridon, *L'imaginaire scientifique*, 192.

ture calls attention to changes in culture and even private life. A hundred years before Philippe Ariès and Georges Duby's large-scale research project into the *History of Private Life*, Viollet-le-Duc was imagining such life in relation to the milieu in which it was lived. The plates in Viollet-le-Duc's *Dictionary of Furniture*, which depict a typical "chambre du seigneur" [castle bedroom] over the course of four centuries, bear witness to this striking shift of attention.[63] They present a combination of historicism and realist genre painting that underscores the novelty of this new historical object.[64]

Along with this shift of historical attention towards the interior, one finds in Viollet-le-Duc's *Dictionary of Furniture* a parallel shift in aesthetic values:

> Pour connaitre une époque, pour prendre une idée quelque peu exacte de ses habitudes, de ses mœurs, il ne suffit point de choisir, parmi les objets qu'elle nous a laissés, certain types rare ou précieux, exceptionnels; il est nécessaire de grouper les objets usuels, les vêtements ordinaires, les armes communes.
>
> To know a period, to have a more or less precise idea of its habits, its mores, it is not enough to select among the objects it has left behind a few rare or precious, exceptional pieces; it is necessary to group together usual objects, ordinary clothing, common arms.[65]

The adjective "usuel" here is both a synonym for "ordinaire" and "commune" as well as an exact description of the principle of categorization: what requires attention are those objects that were *in use* and *of use*. Such objects constitute a milieu because they express or indeed enact the functions of that milieu. As extensions of organs (clothing, arms), such objects are utensils and thus also "organs" in the etymological meaning of the Greek *organon* or "tool." Rejecting aesthetic categories such as the rare, precious, or exceptional, this passage places value in history, or the knowledge of how an epoch was lived.

The utensils of everyday life are the best index for understanding history "as it really was," but this functionalized vision of art objects also redirects the attention of the restorative desire. While the selection of a monument as deserving restoration marks it as special, one can view even exceptional monuments from

---

63 Viollet-le-Duc, *Dictionnaire raisonné du mobilier français de l'époque carlovingienne à la Renaissance*, vol. 1 (Paris: Bance, 1858), plates XII-XV.

64 Baridon concludes his discussion of Viollet-le-Duc's theory of milieu on this note, "Cette mésologie poétique est fondée sur l'image, veritable support de l'imaginaire," [The foundations of this poetic theory of the milieu is the image, the material support of the imaginary] ibid. 201.

65 Viollet-le-Duc, *Dictionnaire raisonné du mobilier français de l'époque carlovingienne à la Renaissance*, vol. 2 (Paris: A. Morel et Cie., 1871), IV; quoted in Baridon, *L'imaginaire scientifique*, 198.

the perspective of their function.[66] Viollet-le-Duc takes this functional approach to architecture in his six novelistic *Histoires*, published as part of Pierre-Jules Hetzel's *Bibliothèque d'éducation et de récréation*, the same series to which Jules Verne's science fiction classics belong. In the *Histoire d'une forteresse*, for example, the architectural development of the fortress is presented through the history of a series of sieges. Military function thus drives architectural form. And if the *Histoire d'une maison* recounts the construction of an "ordinary" bourgeois home, the *Histoire de l'habitation humaine* places "habitation," in the double sense of "building" and "way of life," as the center of its interest. Finally, *Histoire d'un dessinateur*, which Virginia Champlin cleverly translated as *Learning to Draw: Story of a Young Designer* to include both meanings of the French *dessin* ("drawing" and "design"), delivers its idealized young draughtsman into the employ of a furniture manufacturer. Viollet-le-Duc's rehabilitation of the industrial or decorative arts is thus comparable with the contemporary Arts and Crafts movement in England. If, as Martini Bressani writes, "Viollet-le-Duc's dictionary of the Middle Ages is the past put to use: it is future-oriented," then his *Dictionary of Furniture* and *Histoires* establish future construction as a construction of environments.[67]

From Auvergne to Mont-Blanc: Geology as Loss

As we have seen, the priority that Viollet-le-Duc accords to Cuvier's comparative anatomy in the genealogy of restoration corresponds to the special role played by anatomical organicism in his theory of architecture, and by biological theories of milieu in his historiography of habitations, habits, and habitats. Nevertheless, the importance of "geological researches" for restoration, and for the nineteenth-century historical "attitude" more generally, is so far reaching that it almost escapes Viollet-le-Duc's attention in the article "On Restoration." The geological knowledge of reconstructing the past out of the traces that it has left on the present has the status of self-evidence that requires no explanation. Still, the global method of the restoring architect, his "all-important principle" (69), clearly resembles the work of the geologist or archeologist:

---

66 Bressani reads the *Histoires* in two groups, and considers the first group, including *Histoire d'une fortresse* and *Histoire d'un hôtel de ville et d'un cathedrale*, from the perspective of the political history of the Franco-Prussian War and the Paris Commune, *Architecture and the Historical Imagination*, 477. Viollet-le-Duc's functional theory underlies both the historical and transhistorical, "more conservative" *Histoires*.

67 Bressani, *Architecture and the Historical Imagination*, 237.

> Sur ces fragments que l'on ramasse dans des fouilles, il faut examiner les lits de pose, les joints, la taille; car il est telle ciselure qui n'a pu être faite que pour produire un certain effet à une certaine hauteur. Il n'est pas jusqu'à la manière dont ces fragments se sont comportés en tombant, qui ne soit souvent une indication de la place qu'ils occupaient. L'architecte, dans ces cas périlleux de reconstruction de parties d'édifices démolis, doit donc être présent lors des fouilles et les confier à des terrassiers intelligents. En remontant les constructions nouvelles, il doit, autant que faire se peut, replacer ces débris anciens, fussent-ils altérés: c'est une garantie qu'il donne et de la sincérité et de l'exactitude de ses recherches.

> It is necessary to examine the beds (*lits de pose*), joints, and dressing of the fragments collected in the clearing (*fouilles*); for some kinds of tooling could only have been adopted with the view of producing a particular effect at a certain height. The slightest indications, even the way in which these fragments have behaved in falling, may not unfrequently show the place they occupied. In these difficult cases of rebuilding the demolished parts of an edifice, the architect ought, therefore, to be present during the clearing and entrust it to intelligent excavators. In erecting the new constructions, he should as far as possible put these old remains back in place, even if injured: this will furnish a guarantee for the sincerity and the exactitude of his investigations.[68]

During a dig (*fouille*), one finds fragments of the past that indicate a path towards reconstruction. Informing such reconstruction is a geological theory of how the fragments acquired their present form and location (*lits de pose*). Geology appears here as the global science of how objects "behave in falling," or indeed, how things "behave" in being pushed together, pulled apart, driven upwards, downwards, or sideways. During the restoration of the Abbey Church at Vézelay, his breakthrough project, the architect witnessed, surely to his horror, the collapse of an entire vault. At the time, Viollet-le-Duc had already replaced three of the arcades in the main nave, which may have saved the structure as a whole. Little prepared for such engineering tasks by his education as a draughtsman and painter, the young architect's construction acumen at Vézelay remains an important part of his legacy.[69] One imagines that the material investigation of a site would represent an exception, a "difficult case" that arises when architectural elements can no longer by deduced from their own internal, organic logic. Nonetheless, the limit case provides the ground for all others: the ultimate "guarantee" of authenticity and exactness comes from the "hard" science of geology.

---

68 Viollet-le-Duc, "Restauration," 34 [70 – 71]. Translation slightly modified.

69 Bressani, *Architecture and the Historical Imagination*, 100 – 101; Kevin Murphy, *Memory and Modernity: Viollet-Le-Duc at Vézelay* (University Park, PA: Penn State UP, 2000), 108 – 109.

Viollet-le-Duc developed a fascination for mountains and geology at a young age.[70] On trips to the Auvergne in 1831 and to the Pyrenees in 1833, he practiced landscape painting as he confronted the challenges of mountaineering. In his letters, the young Viollet-le-Duc describes this fascination in the typical language of the sublime. Of the Ossau Valley, for example, the nineteen-year-old writes to his father: "Tu ne peux t'imaginer la beauté de l'endroit où je me trouvais [...] peu de personnes aiment et comprennent toute la poésie qu'il y a dans les montagnes" [You can't imagine the beauty of the place where I find myself [...] few people like and understand all the poetry that the mountains have to offer].[71] The sublimely beautiful mountains also remind the young man of his own limitations:

> La comparaison de l'abondance, de la puissance des moyens de cette nature sauvage, son calme même écrase la faiblesse des moyens de l'homme, son activité mesquine et son audace présomptueuse.
>
> The comparison of the abundance and power of this savage nature, even its calm crushes the weakness of human means, their paltry endeavors and presumptuous audacity.[72]

In a watercolor of the "Bottom of the Azun Valley taken from the Pourgues Mountain, Hautes-Pyrénées," painted in 1833, Viollet-le-Duc translates this insignificance into the visual language made famous by Caspar David Friedrich: a wanderer reclines in the lower right corner of Viollet-le-Duc's composition, the line of his back almost disappearing into the incline of the foregrounded slope, while the center of the painting and its vanishing point lead the eye to majestic, snow-capped peaks in the background.

In addition to these almost clichéd citations of the (visual) language of the sublime, the young Viollet-le-Duc also demonstrates an interest for an aspect of geology that resonates strongly with his later career as a restorator. A number of his early sketches focus on scenes of rubble and erratic blocks, which he describes in a letter in terms that highlight the destructive moment of sublime contemplation:

---

70 Cf. Christiane Amiel, "Viollet-le-Duc et la montagne," in *Eugène Viollet-le-Duc: regards croisés. Actes du colloque de Carcassonne 5 et 6 juin 2014*, ed. Jean-Pierre Piniès (Carcassonne: Garae Hésiode, 2014), and the essays and prints in the exhibition catalogue Frey and Grenier, eds., *Viollet-le-Duc et la montagne.*

71 Cited in Geneviève Viollet-le-Duc, *Les Viollet-le-Duc. Histoire d'une famille. Documents et correspondance* (Geneva: Slatkine, 2000), 167–168.

72 Ibid. 171.

> Autour ce n'est que débris, neiges et forêts; des blocs de granit percent au milieu des sapins, tout cela sent la destruction, les ravages, la mort. C'est horriblement triste, mais quelles lignes admirables [...] Je me trouve dans mon élément, au milieu de ces grandes et belles horreurs que j'aime tant.
>
> All around there is only rubble, snow, and forests; blocks of granite erupt in the midst of fir trees, all of this speaks of destruction, devastation, death. It is horribly sad, but what beautiful lines. [...] I found myself at home in the midst of these beautiful horrors I love so much.[73]

It is worth noting that, only two years prior, Viollet-le-Duc had written at the top of a drawing of the Puy-du-Dôme, a volcanic peak in the Auvergne, "Il n'y a que les montagnes qui ne bougent pas" [Only mountains do not move].[74] While the intervening death of the young man's mother likely played a significant role in his change of mood, his morbid fascination for the destructive side of nature corresponds with the advances in geological knowledge of his time. Indeed, in a semantic amalgam that captures both his personal mourning and a broader geological insight, he names the force behind this destruction a "*puissance mère*" [mother force].[75]

Such powerful personal experiences in the mountains could have motivated a career as a geologist and Alpinist. Indeed, Viollet-le-Duc's return to geological pursuits at the end of his life suggests that the passion never left him. There is a shift, however, in Viollet-le-Duc's attitude towards mountains that occurs during his travels in Italy in 1836. From the summit of Mount Etna in Sicily, Viollet-le-Duc, now married, still writes to his young wife of the thrills of mountaineering, "il y a toujours dans les ascensions quelque chose qui fouette le sang" [there's always something to a climb that whips the blood].[76] By the time of his return crossing of the Alps in 1837, however, he registers, with regret, that the mountains no longer hold sway over his imagination, "Non, plus rien, un sentiment de curiosité, voilà tout. Cela est horriblement triste, je te jure, d'être ainsi déçu..." [No, nothing, a feeling of curiosity, nothing more. It's horribly sad, I swear to you, to be disappointed like this...].[77] Although this experience ignited

73 Viollet-le-Duc, letter to his father from Pau, June 28, 1833, in *Les Viollet-le-Duc*, ed. Geneviève Viollet-le-Duc, 172, and from Gavarnie, July 18, 1833, ibid. 189. Translation by Bressani, *Architecture and the Historical Imagination*, 74.

74 Reproduction in Lisa Grenier, "Il n'y a que les montagnes qui ne bougent pas," in *Viollet-le-Duc et la montagne*, ed. Frey and Grenier, 12.

75 Cited in Bressani, *Architecture and the Historical Imagination*, 75.

76 Viollet-le-Duc, *Lettres d'Italie (1836–1837) adressées à sa famille*, ed. Geneviève Viollet-le-Duc (Paris: Léonce Laget, 1971), 85.

77 Ibid. 338–339.

his passion for painting, art, and architecture, it also left him relatively cold to the beauty of nature.

As a series of remarkable images from his Italian journey demonstrate, Viollet-le-Duc's geological interests did not so much disappear as they were absorbed into his nascent restorative desire. In Martin Bressani's careful reconstruction of this process, these images constitute a virtual trajectory from the mountain rubble of the Pyrenees, to the debris surrounding the "Temple of Junon Lecine at Agrigento" (1836), and finally, to two remarkable images depicting the Roman theater at Taormina, with Mount Etna in the background.[78] Along this path, Viollet-le-Duc both associates and then progressively replaces images of natural rubble with cultural ruin. He also replaces the "beautiful horror" of the sublime with a new sentiment: "Il n'y a pas de sensations plus vives et plus profondes que celles que l'on éprouve en se trouvant au milieu de tous ces vénérables débris. Peu à peu l'imagination restaure des énormes murailles d'un seul morceau, ... [et] bientôt on peuple tout cela" [There are no livelier and more profound sensations than that felt when standing among these venerable rubbles. Bit by bit the imagination restores these enormous monolithic walls, [...] and soon the mind fills the place with people].[79] Viollet-le-Duc repeats such imaginative flights of fancy at the Coliseum and the Baths of Caracalla – which he depicts in drawings that "accentuate their geological nature, distilling the buildings down to their brutal mass and the sheer strength of the rock thanks to a brown monochrome wash and very sharp shadows."[80]

While Viollet-le-Duc's youthful sketches depict mountains and ruins in the same visual vocabulary, his depictions at Taormina on the eastern coast of Sicily bring these two phenomena together and establish a practice of restoration based on this link. A drawing made during the trip in 1836 views Mount Etna from the town of Taormina, with the ruins of an ancient Greek theater in the foreground. In a monochrome treatment similar to that of his drawings at Agrigento and Caracalla, this sketch establishes a continuity between the ruined amphitheater and the crater-form of the volcano, and thus between the destructive geological forces that created these forms. Deeply moved by this particular composition, the aspiring artist proclaimed it a view "unmatched in France, Naples, or anywhere else."[81] Viollet-le-Duc also chose it as the motif for a historical paint-

---

78 Bressani, *Architecture and the Historical Imagination*, 73–81.

79 Eugène-Emmanuel Viollet-le-Duc, *Lettres d'Italie*, 70. Translation by Bressani, *Architecture and the Historical Imagination*, 79.

80 Bressani, *Architecture and the Historical Imagination*, 79.

81 Viollet-le-Duc, *Lettres d'Italie*, 87. Translation by Bressani, *Architecture and the Historical Imagination*, 7.

ing that he later exhibited at the Salon of 1840 in Paris. As Bressani notes, the historical reconstruction of ruins in the virtual medium of historical painting is a form of "restoration" popularized in the 1830s by painters known as the *pensionnaires* at the Académie de France in Rome.[82] This genre of historical painting inspired Viollet-le-Duc to undertake his first restoration project, the Greek theater at Taormina, even if it was never carried out on the building itself.

What stands out in the later picture, however, is not only the vibrant Greek theater filled with spectators, but also the smoking Etna looming in the background. In reviving the theater, Viollet-le-Duc also revives the threat of destruction represented by the active volcano. The "complete state" that the painting depicts may be ahistorical, in the sense that it may never before have existed, but it does not exist outside of history, or at least, not outside of earth history. On stage, the Furies gather as the tragedy of Aeschylus's *Eumenides* unfolds, but the background is where the catastrophe of geology takes place. Etna leaves as indelible a mark as any ruins Viollet-le-Duc visited in Sicily: "Le gouffre ... a cet attrait si puissant pour l'imagination [...] parce qu'il est vraiment beau et au-dessus de ce qu'on peut se figurer. Ses grandes et longues crevasses fumantes, ses pointes hardies et déchiquetées ne me sortent pas de la mémoire" [The abyss is a powerful attraction to the imagination [...] because it is truly beautiful and so much above what one can conceive. These great, long smoking crevasses, these bold, jagged peaks, cannot leave my memory].[83] Geology thus remains present in the young man's mind as the memory of an abyss, as the memory of destruction and loss.

Despite or because of this negative form of memory, Viollet-le-Duc's return to Paris after his Italian trip represents the beginning of a long caesura in his career as a mountaineer and geologist. Concentrating his efforts on architecture and restoration, beginning with the imaginary restoration of the theater at Taormina, Viollet-le-Duc only returned to the mountains regularly in the late 1860s. His increasingly extensive geological studies in the Swiss Alps and especially in the Mont-Blanc region, however, soon resulted in an extensive geological study of the Mont Blanc. He published this work in 1876 under the title *Le massif du Mont Blanc, étude sur sa constitution géodésique et géologique, sur ses transformations et sur l'état ancien et moderne de ses glaciers* [*Mont Blanc. A Treatise on its Geodesical and Geological Constitution; its Transformations; and the Ancient and Recent State of its Glaciers*]. As the title suggests, Viollet-le-Duc's geo-

82 Bressani, *Architecture and the Historical Imagination*, 7.

83 Viollet-le-Duc, *Lettres d'Italie*, 84. Translation (slightly modified) by Bressani, *Architecture and the Historical Imagination*, 80.

logical knowledge finds its place between geodesy and glaciation. If geodesy relies on geometric forms to imagine the ancient or original state of mountains, glaciation represents the erosive forces that transform a mountain range to its present state. Linking the two domains is the image of the crystal and the crystalline forms of mountain granite: "des lois ont ordonné ces formes, ont déterminé le grand système cristallin" [laws have put these forms in order and determined the structure of the grand crystalline system].[84]

Because he imagines the formation of mountains and rock primarily as a process of crystallization, Viollet-le-Duc employs various pyramids and rhomboids to model their original state, though his preferred form is the rhombohedron (cf. *Le massif du Mont Blanc*, Fig. 47). As Jan von Brevern has demonstrated, this method of geological reconstruction is inseparable from the practice of drawing that underlies it.[85] When Viollet-le-Duc sketches mountain formations, much as he learned in his youth, he seeks the abstract geometrical forms behind the visible surface. Viollet-le-Duc thus practices geology in much the same way as he practices architectural restoration, a connection he makes explicit in his preface to *Le massif du Mont Blanc:*

> De fait, notre globe n'est qu'un grand édifice dont toutes les parties sont coordonnées et ont une raison d'être; sa surface affecte des formes commandées par des lois impérieuses et suivis d'après un ordre logique.
>
> Analyser curieusement un groupe de montagnes, leur mode de formation et les causes de leur ruine; reconnaitre l'ordre qui a présidé à leur soulèvement, les conditions de leur résistance et de leur durée au milieu des agents atmosphériques; noter la chronologie de leur histoire, c'est, sur une plus grande échelle, se livrer à un travail méthodique d'analyse analogue à celui auquel s'astreint l'architecte praticien et archéologue qui établit ses déductions d'après l'étude des monuments.

> In fact, our globe is but a grand edifice whose every part has a reason for being; its surface displays forms ordered by imperious laws and demanded by a logical order.
>
> To analyze with care a group of mountains, the way they were formed and the cause of their ruin; to recognize the order that presided over their elevation, the conditions of their resistance and durability in the midst of atmospheric agents; to document the chronology of their history: it is, on a larger scale, to carry out a methodical work of analysis analogous to that carried out by the architect-archeologist who establishes his deductions upon the study of monuments.[86]

84 Viollet-le-Duc, *Le Massif du Mont Blanc,* 240.
85 Jan von Brevern, *Blicke von Nirgendwo,* 102–108; cf. Bressani, *Architecture and the Historical Imagination*, 485–491.
86 Viollet-le-Duc, *Le Massif du Mont Blanc*, xv-xvi.

In Viollet-le-Duc's analogy, the difference between geological and archeological research is only a matter of scale. In his earlier work, Viollet-le-Duc relied on an organicist analogy to transfer Cuvier's geological knowledge to the realm of architecture. Now, he returns his geohistorical theory of architecture to its original domain. As plausible as this re-entry sounds in theory, the idea that the globe is a "vast edifice" confounded contemporary geologists. The president of the *French Geological Society*, Alphonse Favre, encountered Viollet-le-Duc on an excursion or "réunion extraordinaire" of the society in 1875. Favre summarizes Viollet-le-Duc's claims with understated befuddlement, "There is, according to this scholar, a close association between certain aspects of geology and architecture; Mont Blanc is a ruin; one can find its original form by following the same principles that one uses in restoring a monument."[87] A less restrained critic of Viollet-le-Duc's geological work, John Ruskin, complained of its "provincial temper" and "insolent errors," and suggested its goal was "the architectural restoration of the Gothic globe."[88]

Rational to a fault, Viollet-le-Duc's method of geological research more resembles the universal geological theories of the eighteenth and early nineteenth centuries than the complex geological history that became scientific consensus over his lifetime. As the quotation above makes clear, the architect imagines the "composition" of geological formations as following a strict order. Reconstruction is thus a "work of deduction" because the processes of decomposition are secondary. The architect's geological reconstructions resemble blueprints that reinstate an original, geometrical form over the eroded, observable surface.

The close relationship between geology and (architectural) drawing is also evident in Viollet-le-Duc's final work, *Histoire d'un dessinateur.* As the French subtitle suggests – *Comment on apprend à dessiner*, which translator Virginia Champlin sets as the primary title *Learning to Draw* (1881), – this short novel of education also functions as a drawing manual for children. The final drawing lesson, "Twelve Days in the Alps," follows the young boy Jean and his mentor M. Majorin on a trip through Switzerland. Concluding Jean's aesthetic education, the chapter represents a narrative climax, pedagogical summa, and aesthetic highlight of the book. At the same time, the most difficult lessons in Viollet-le-Duc's drawing curriculum are quite simple when compared to the complexity of the object to be drawn. Majorin's lessons underscore the fundamental similarity between geological drawing and the earlier lessons in anatomy, which, both

87 Alphonse Favre, "Réunion extraordinaire," *Bulletin de la Société géologique de France* 3 (1875); quoted in Jan von Brevern: *Blicke von Nirgendwo*, 93.
88 John Ruskin, *The Works of John Ruskin*, vol. 26, 224.

because of their traditional role in drawing and the importance that Viollet-le-Duc ascribes to comparative anatomy, occupy two key chapters at the start of the novel. Before embarking on an extended explanation of glacial erosion and speculations on the cooling earth, which constitute the chapter's geological lessons, Majorin thus admonishes his pupil to remind him of the most important rule:

> "Oh! mais," lui dit M. Majorin pendant qu'il ébauchait son croquis, "tu en prends à ton aise avec ces roches; cela demande à être dessiné avec précision, et ces formes ne sont pas dues au hasard. Il y a de grandes lignes principales qu'il faut d'abord indiquer; puis, quand elles sont fidèlement tracées, en tenant compte des angles et des inclinaisons, il faut y faire entre les détails suivant leur importance [...] chaque nature de roche affecte des formes spéciales, et il ne faut pas plus les négliger qu'on ne néglige d'observer les traits d'un visage ou les allures d'une plante."
>
> "Oh stop!" said M. Majorin, while Jean was sketching: "you are too free with those rocks, which require to be drawn with great exactness, for their forms are not due to chance. There are prominent principle lines which you must first indicate; then, when they are faithfully outlined, omitting none of the angles and inclinations, you must fill in the details according to their importance. [...] each kind of rock has its own form, and it must be no more neglected than the features of a face or the lines of a plant."[89]

Viollet-le-Duc's geological drawing method thus begins with a schema, and then fills in the details. The method relies on a fundamental analogy between the laws that create the forms of rocks and the forms of human and plant anatomy. Of course, geological science is obliged to explain how this schema came to resemble the empirical reality one can see and observe. For this reason, Majorin also insists "rien n'est à négliger dans l'étude des montagnes" [in the study of mountains every thing is important]. Only when the young Jean follows this advice, does he learn, through careful drawing that "il y a dans leur forme, et jusque dans la façon dont elles se ruinent, des lois générales; que telle disposition conduit invariablement a tel fait" [there are general laws governing their forms, and even the manner in which they crumble away, and that a certain arrangement invariably leads to a certain fact].[90] Through the voice of Majorin, Viollet-le-Duc thus attempts to give his drawing lessons a solid scientific foundation. In this context, it is notable that Viollet-le-Duc does not reproduce any of his own geometrical "restorations" of mountain forms in *Learning to Draw.* Instead,

89 Viollet-le-Duc, *Histoire d'un dessinateur: comment on apprend à dessiner* (Paris: J. Hetzel, 1879), 255–256; *Learning to Draw: Or the Story of a Young Designer,* trans. Virginia Champlin (New York: G.P. Putnam's Sons, 1881), 273.
90 Ibid. 261 [278].

the pedagogical text focuses on either conventional landscape sketches (fig. 95, 100, 104), or diagrams that illustrate geological laws such as glacial valley formation (fig. 96, 98), glacial friction (fig. 97), upheaval (fig. 99) and the cooling of the earth's crust (fig. 101, 102).

Like the article on "Restoration," *Learning to Draw* is a retrospective reflection on the author's work. Viollet-le-Duc's novel of education includes several autobiographical clues that allow informed readers to associate Jean with his younger self. In the figure of M. Majorin, however, one can identify not only Viollet-le-Duc's uncle and mentor Etienne Delécluze, but given the content of his discourse, the author himself. As an idealized reimagining of the author's own education, *Learning to Draw* thus strongly resembles the structure of *Indian Summer*, and like Stifter's novel, it can be read as a kind of narrative "self-restoration."[91] The autobiographical reading of Viollet-le-Duc's final text becomes particularly interesting in light of his discussion of geological phenomena. For example, the aging author describes fictional Jean's fascination with the mountains in terms that recall the letters of his youth, but also dismisses them as naive: "Inutile d'énumérer les étonnements et les exclamations de petit Jean pendant ce trajet" [It is useless to repeat Jean's exclamations of astonishment during this trip].[92] In place of the aesthetic that describes the mountain landscapes as "affreux" [frightful] and "désolé" [desolate],[93] *Learning to Draw* focuses on the second moment of the sublime feeling, the knowledge that leads to an enjoyment of and passion for the mountains that is "guère compris" [rarely understood], though shared by pupil and mentor alike.[94] Indeed, without geological knowledge, there would be no interest in mountaineering:

> Ce n'est que depuis la science est venue révéler à l'homme quelques-uns des mystères de la nature, que l'on s'est mis à parcourir les montagnes avec le désir de les connaitre et quelques-uns se sont pris de passion pour les spectacles grandioses des altitudes. Les hommes primitifs redoutaient ces hauteurs et les considéraient volontiers comme le séjour des divinités, qui n'étaient à tout prendre que la personnification des forces de la nature.
>
> It was not until science revealed to man the mysteries of nature, that he began to travel through the mountains with a desire to know them, and that he acquired a passion for the grand views from mountainous heights. Primitive man feared them, and looked

91 Bressani, *Architecture and the Historical Imagination*, 499. Bressani also deciphers the autobiographical details of this *roman à clef*, 502.
92 Viollet-le-Duc, *Histoire d'un dessinateur*, 249 [265].
93 Ibid. 252 [269].
94 Ibid. 252–253 [269–270].

> upon them as the abode of divinities, who were, after all, only personifications of the forces of nature.[95]

While Majorin contrasts modern science with the fears of "primitive man," his own understanding of geological processes creates a new fear. For in the geological imagination of this late text by Viollet-le-Duc, the "forces of nature" in the mountains are primarily destructive: "De ces sommets partaient les tempêtes, descendaient les torrents destructeurs et les fleuves fertilisants" [From these summits tempests arose, and destructive torrents and fertilizing rivers descended].[96] Indeed, the fact that Majorin considers the mountain erosion as the source of fertile soil for the plains leads to the most disturbing vision. As the summits erode and the rivers sweep away their debris into the valleys, "some day only scattered ruins of these lofty mountains will remain." Lacking elevation, the glaciers themselves will melt and disappear, and without a source of nourishing soil and life-giving water, the plains will become a "steppe, marécageuse ou aride" [swampy or arid steppe]. The final consequence of this universal process of erosion and desertification is mass extinction: "Et alors les races humaines, les animaux disparaitront peu à peu de la surface de la terre faute de pouvoir y vivre, comme elles ont disparus sur quelques points de notre globe ou se sont produits les phénomènes analogues depuis les temps historiques, c'est-à-dire depuis hier" [The human race and animals will gradually disappear from the earth, because it will be unable to support life, as they have disappeared on several parts of the globe where, within the historical period, that is, since yesterday, similar phenomena have occurred].[97]

While the pessimism of this vision may owe to Viollet-le-Duc's advanced age, its logic emerges from an experience of geological time that had been accessible to him since his first youthful adventures in the Pyrenees. The knowledge that the "historical period" is subject to the same geological laws as the deep prehistory of the planet provokes the vision of a world without humanity. The work of drawing, reconstructing, and restoring follows as a reaction to the shock of this perceived loss. At the end of his lessons in drawing, the young Jean prepares himself for the Central School where he will learn to design (the second meaning of *dessiner*). After his trips to the mountains, he is thus prepared not only for the most difficult form of drawing, the live model, but also to begin making models ("il dut dessiner d'après le modèle vivant, et il se mit à modeler"). The telos of

95 Ibid. 253 [270].
96 Ibid.
97 Ibid. 255 [272]. Translation slightly modified.

Viollet-le-Duc's art is thus to model life by reproducing its form, or as Jean's teacher Majorin concludes: "celui qui sait dessiner doit être en état de modeler, et quand on conçoit bien une forme, il n'est pas plus difficile de la rendre à l'aide de l'ébauchoir qu'à l'aide du crayon; il faut donc s'habituer à se servir indifféremment de l'un ou de l'autre" [Whoever knows how to draw must be able to model; and, when one perfectly conceives a form, it is no more difficult to render it by the aid of the chisel than by a pencil: one must therefore accustom himself to freely use either].[98] In Viollet-le-Duc's geological imagination, his pencil is always already a chisel, working the stone of castles and cathedrals just as the glacier carves away beautiful forms in a mountain valley. In the transition from pencil to chisel, from project to realization, however, the young designer finally encounters stone as a material limit to his creative genius. The glacier produces beautiful forms, until all that is left is rubble.

Kindred Spirits: Viollet-le-Duc and Stifter

This horizon of geological obscurity and fundamental instability links the French architect Viollet-le-Duc and the Austrian writer Stifter. At the margins of their respective theory and practice of restoration or *Wiederherstellung*, there is a disconcerting knowledge of the destructive forces of earth history. In the case of Viollet-le-Duc, this knowledge remains marginal in a biographical sense. His fascination with the Alps and the mountain sublime is strongest at the start and end of his career, while in the interval, he strives to counteract geological forces of erosion through architectural restoration. As we saw in the last chapter, Stifter's novel *Indian Summer* integrates the disturbing impact of geology into its poetic structure, first as a past that can never be recovered, then through a disruptive emergence of the present tense in the young geologist's anxious contemplation of an uncertain future.

Each author finds a memorable phrase to describe the ontological precariousness of their respective restorative poetics. As Stifter speaks of a "post-summer without a preceding summer," Viollet-le-Duc envisions reestablishing a "complete state that may never before have existed." Both standpoints are modern in the sense that they ground the understanding of their own place in history in distinction to what came before. At the same time, their restorative historicism seeks to recover a past that it paradoxically acknowledges as irretrievable or even nonexistent. These definitions recognize the fictionality of their own constructions of

98 Ibid. 273 [291].

the past. In this sense, they go beyond the romantic-nationalist fascination with the Gothic or Middle Ages, with *ancienne France* and *altdeutscher Kunst.* While such interest continues to motivate both Stifter and Viollet-le-Duc, they distance themselves from forms of direct revival. Indeed, by acknowledging the unprecedented modernity of their practice, these theoretical self-reflections on restoration also approach a postmodern understanding of history. It is no surprise, at least, that the rise of postmodern theories in the late twentieth century led to the first positive reappraisals of their work.

Beyond this deep discursive affinity, there are a remarkable number of smaller similarities between these two geological historicists. Both not only recognize the historicity of their own historicist practices, but they also concur in the dating of restoration to the second quarter of the nineteenth century. In *Indian Summer*, we see through the eyes of a geologist how the objects of the history of art might have a systematic logic: "Da ich nun alle diese Laubwerke diese Ranken diese Zacken diese Schwingungen diese Schnecken in großer Abfolge sah, erschienen sie mir gewissermassen wie Naturdinge etwa wie eine Pflanzenwelt mit ihren zugehörigen Thieren" [As I now looked at all this foliage, these tendrils, points, curved lines, spirals in great succession, they seemed to me like natural objects, a plant kingdom with its appropriate animals].[99] In the person of Viollet-le-Duc, one finds the historical reality of an art historian or archeologist whom the fictional Drendorf imagines speculatively: "man könnte sie eben so zu einem Gegenstande der Betrachtung und der Forschung machen wie die wirklichen Pflanzen und andere Hervorbringungen der Erde, wenn sie hier auch nur eine steinerne Welt sind" [a person could make them the subject of observation and research just as well as the real plants and other things of the Earth, even if they are only a world in stone].[100] While the fictional geologist thus registers the organicist analogy with a skeptical concession, "even if they are only," the phrase "world in stone" reinforces the poetic imaginary that so inspired Viollet-le-Duc, who loved the mountains and ruins equally, as stones.

The French architect, of course, not only systematizes historical objects in a "tableau," but also has a firm faith in the predictability of the line of filiation, or the *Abfolge*, as a stylistic evolution that can even serve as a guide for reconstructing 'missing links.' Or as Stifter writes:

> Diese Dinge haben so gut wie bedeutendere Gegenstände ihre Geschichte, und aus dieser Geschichte kann man das Aussehen und den Bau derselben zusammen sezen. Im Verlaufe der Jahre habe sich die Gestaltungen der Geräte immer neu abgelöset, und wenn man auf

99 HKG 4.1, 104 [63].
100 Ibid.

> diese Abfolge sein Augenmerk richtet, so kann man aus dem vorhandenen Ganzen auf verloren gegangene Theile schließen, und aus aufgefundenen Theilen auf das Ganze gelangen.
>
> These things, like more significant objects, have their own history, and from this you can deduce the appearance and construction. In the course of the years the shape and forms changed predictably, and if you pay close attention to the sequences (*Abfolge*), from an available whole you can deduce lost pieces and reconstruct the whole from what is extant.[101]

This recalls the relationship of furniture to architecture, and in particular, Risach's statement "daß die Bauwerke einer Zeit und die Geräte, welche in diesen Bauwerken sein sollten, eine Einheit bilden, die nicht zerrissen werden kann" [that buildings and the appropriate furniture of a given period form a unity that should not be torn asunder].[102] Changing forms result from changing "styles," and the true expression of style is an architectural ensemble, interior and exterior, in other words, a *milieu*. Though the Freiherr von Risach's own museum-like house more resembles the eclectic historicism typical of bourgeois residences, he aids his friend Mathilde in furnishing her palace, the Sternenhof, according to this principle: each room is the reconstructed milieu of an epoch in the history of the palace. In other words, the Sternenhof is furnished in the same way as the imagined reconstructions that illustrate Viollet-le-Duc's *Dictionary of French Furniture.*

In Stifter's novel, the difference between Mathilde's palace as a historical monument and the Risach's more contemporary home derives from the function of this home. It serves as the site of a kind of modern corporation or foundation, an *oikos* that combines agriculture with cultural conservation and production. The provincialism apparent in this vision of a rural cultural center, that is, an eccentric center, corresponds to the decentralization that Viollet-le-Duc imagined as a healthy counterbalance to the driving force of centralization in the politics and economics of nineteenth-century France. Both authors imagine the provinces as the primary beneficiaries of a cultural revival set in motion by the work of conservation. Viollet-le-Duc dreams of architects and artisans who, in learning the techniques of the past, reach a new level of artistic achievement: "C'est grâce à ces travaux de restauration, que des industries importantes se sont relevées" [It is owing to these enterprises in restoration that important branches of industry have been revived].[103] Masonry, glass and ironwork, and many other branches of the construction industry benefit from the challenge of restoration.

101 HKG 4.1 98 [60].

102 HKG 4.1 108 [65].

103 Viollet-le-Duc, "Restauration," 28 [53].

Stifter fictionalizes such success stories in the characters of Eustach, a local artisan who draws on his historical knowledge to design new pieces of furniture, and his brother Roland, who as a painter transfers such knowledge to the realm of high art.

As Viollet-le-Duc notes, the provincialism of historical conservation work is a pragmatic consequence of the geographical distribution of historic monuments. Many of the specific positions on questions of conservation practice – and the similarities and differences between them – are intertwined with the local situations of the authors, and above all their institutional affiliations in Upper Austria and France, respectively. The goal of this excursion, however, was not so much to produce a detailed historical comparison of restoration practices in Austria and France in the nineteenth century, so much as to examine the way in which two practitioners understood the genealogy and historical underpinnings of their own work.

As I have shown, geology plays a central role in this genealogy. It provides both a motive and a method for a practice of historical reconstruction focused on physical materials and milieus of the past. Like other discourses of modernity, but more radically because it concerns the whole of nature, geology motivates a desire for reconstruction because it disrupts the continuity between present and past, and thus constructs the past as loss. At the same time, geology provides knowledge and method for recognizing the *traces* of this past. Indeed, everything, the whole of the *nature and culture*, becomes a trace of what was.

The analogy between nature and culture, a defining feature of the discourse of geohistoricism, ultimately relies on a concept of "trace" that correlates to the art of *drawing* and realizes the analogy in practice. In *Indian Summer*, Heinrich Drendorf first records the traces of earth history with his pencil, then, having discovered the archive of architectural drawings in Risach's studio, turns his attention to the traces of the cultural past as well. Drawing and painting are so central to Drendorf's educational story that Stifter's novel could bear the title of Viollet-le-Duc's final work: *Histoire d'un dessinateur.* An autobiographical *Bildungsroman* for children, this text not only seals the French architect's legacy as one of the finest draughtsmen of the nineteenth century, but also demonstrates the author's self-awareness of the centrality of drawing in his work as an architect, restorator, and geologist.

The actual drawings or illustrations in *Histoire d'un dessinateur* point to a divergence in the way the two authors present this self-awareness. In Viollet-le-Duc's *Learning to Draw*, the illustrations are the primary medium of the work's didactic message. While the narrative provides an entertaining context – or rather, pretext – for presenting a series of didactic drawings, the author has drawn them so that his young readers can imitate and reproduce them, as in any other drawing manual. At the same time, in this and other works by the architect and geologist, drawing has an epistemological value. Viollet-le-Duc's images suggest that the processes of mountain formation, erosion, or architectural evolution occurred in the way that the drawing illustrates. By contrast, Stifter's literary descriptions of drawing practices place them in a different light. Most importantly, the novel reflects upon, but does not enact the analogy between geology and architecture. A difference remains between the "world of stone" that Drendorf sketches and maps in the mountains and the one he discovers in the albums of the restoration studio.

While Stifter wrote *Indian Summer*, however, he also imagined a more direct literary form of historical reconstruction. Inspired by both Walter Scott's popular historical novels and philologically reconstructed medieval epics, Stifter eventually transformed years of research on the history of Bohemia into the historical novel *Witiko*. With its promise to represent lived history, the genre of the historical novel provides the most direct literary equivalent of the restorations in stone and painting imagined by Viollet-le-Duc. Like the architect's fanciful restorations, the historical novel also blends fact and fiction, which leaves its author to decide how to frame this distinction. Relying on geohistorical practices of reconstruction, Stifter makes a strong claim to historical veracity in *Witiko*, and more importantly, as we shall see in the next chapter, to the vitality of its cultural-historical depictions.

Chapter 4
Traces of Life: The Limits of Geohistoricism

Memento vivere: History and Life

One critic of historicist culture, Friedrich Nietzsche, was as convinced as Viollet-le-Duc that the historical outlook of his contemporaries was unique. In the introduction to this study, Nietzsche's critical take on "world history" provided a useful framework for understanding the impact of geological "deep time" on nineteenth-century culture. In its conclusion, the philosopher's reflections "On the Utility and Liability of History for Life" can help frame a critique of the internal contradictions of geohistoricism.[1]

Nietzsche's intent in the second of his *Unzeitgemäße Betrachtungen* [*Unfashionable Observations*], "Vom Nutzen und Nachtheil der Historie für das Leben" [Utility and Liability of History for Life, 1874], does not, at first glance, have much in common with the historical conservation movement.[2] In fact, the two positions seem at odds with each other. While historical conservationists aim to preserve the past they venerate, Nietzsche appears to see such excessive interest in history as a symptom of his age's "*historisch[e] Krankheit*" [historical sickness], adding: "es giebt einen Grad, Historie zu treiben und eine Schätzung derselben, bei der das Leben verkümmert und entartet" [there is a way of practicing history and a valorization of history in which life atrophies and degenerates].[3] Although Nietzsche's essay repeatedly and vigorously rejects every "Uebermaass

1 Katharina Grätz, *Musealer Historismus*, 72–77, discusses Nietzsche as the most prominent critic of historicist culture in the nineteenth century and presents a parallel to Stifter's ironic depicition of a (monumental) historical obsession in the novella *Tourmaline*, 67–72. In both authors she identifies excessive collection as a cause of information overload.

2 As Ernst Schulin shows, Nietzsche's critique does not apply directly to the work of contemporary historians either, "Zeitgemäße Historie um 1870. Zu Nietzsche, Burckhardt und zum 'Historismus,'" *Historische Zeitschrift* 281 (2005): 34–42. As I argue, Nietzsche's *Historie* is more general, and designates a broad 'historical turn' behind conservationism, the historicism of historians, but also the historicization of nature, language, etc. and of course, the earth.

3 Friedrich Nietzsche, "Unzeitgemäße Betrachtungen II. Vom Nutzen und Nachtheil der Historie für das Leben," in *Die Geburt der Tragödie; Unzeitgemäße Betrachtungen I–IV; Nachgelassene Schriften 1870–1873*, ed. Giorgio Colli and Mazzino Montinari, Kritische Studienausgabe 1 (Munich, Berlin, New York: DTV/De Gruyter, 1988), 329, 245. Friedrich Nietzsche, "On the Utility and Liability of History for Life," in *Unfashionable Observations*, trans. Richard T. Gray, The Collected Works of Friedrich Nietzsche 2 (Stanford: Stanford UP, 1995), 163, 85. In the following, page numbers in brackets refer to this translation.

https://doi.org/10.1515/9783110685534-006

des Historischen" [excess of history], his essay would likely not be read today as a classic if it were merely a polemic.[4]

Indeed, the famous sections of Nietzsche's essay are the ones at its beginning and end, which the philosopher describes as having a "Ruhe" [calm tenor], which contrasts with the more bitter diatribes of its middle.[5] These polemical interceding chapters address aspects of "historische Bildung" [historical cultivation] that are exclusively negative in Nietzsche's estimation: the historian's passive objectivity, bigoted presentism, the self-satisfaction of Hegelian teleology, or the information overload of historicist positivism.[6] By contrast, the opening chapters discuss three ways of approaching history, the "monumental," the "antiquarian," and the "critical," which present their own liabilities, but also have their own utility for life. In a similar vein, the last chapter praises "das Unhistorische und das Ueberhistorische [als] die natürlichen Gegenmittel gegen die Ueberwucherung des Lebens durch das Historische, gegen die historische Krankheit" [the ahistorical and the suprahistorical [as] the natural antidotes to the stifling of life by the historical, to the historical sickness].[7] While these substances may be poisonous on their own, they are effective palliatives when dosed correctly.

Two aspects of Nietzsche's critique of history apply specifically to the problems of geohistoricism. First, along with contemporaries like Viollet-le-Duc, who are proud of their century's novel approach to history and earth history, Nietzsche sees his own period as eminently historical. The attitude he describes as "antiquarian," moreover, closely aligns with the aesthetic practices and preferences of pioneering historical conservationists such as Stifter and Viollet-le-Duc. If Nietzsche provides a philosophical vocabulary for reevaluating the aims of historicism, Stifter can once again provide an example of the discourse's practical dilemmas, in particular, when historical restoration is applied to literary production. For both discussions, however, the key term is "life," which appears as a necessary counterpart to a form of historicism that finds its theoretical and practical grounding in the concept of milieu. As Georges Canguilhem establishes in his now classic essay, the milieu or environment is always relative to the living being that inhabits it.[8] For Nietzsche, the driving question is to what extent "history" contributes to a healthy milieu for present and future life. Such a question challenges the "viability" of historical conservationism in both the sense of

4 Ibid. 329 [163].
5 Ibid. 330 [163].
6 Ibid. 246 [86].
7 Ibid. 331 [164].
8 Georges Canguilhem, "Le vivant et son milieu," 165–198.

its desirability and its feasibility. As my reading of Stifter's historical novel *Witiko* demonstrates, the vexing question of "historical life" goes to the heart of the contradictions inherent in the aesthetic practice of geohistorical reconstruction.

Nietzsche's "Concept-Quake"

Before applying Nietzsche's specific critique of the "antiquarian" as it applies to the geohistorical, however, it is worth recalling the main points of his argument as a whole. Nietzsche evaluates all modes of historical practice with respect to how they serve the interests of "life." Life is a central concept in Nietzsche's philosophy, closely related to "power," a value that, one can argue, is itself grounded in biological necessity.[9] Within the context of his essay on history, "life" has two fundamental characteristics: it is bound to the present; and it resides in the subjective needs of the living. Insofar as it deals with the past, history is prima facie antithetical to present life, or at least, the "atmosphere" of living beings – a metaphor that I will examine in detail below – is fundamentally "ahistorical."[10] Second, one can deduce that life is "subjective" from the fact that the academic or "scientific" approach to history is the most antithetical to life: "Die Geschichte als reine Wissenschaft gedacht und souverän geworden, wäre eine Art von Lebens-Abschluss und Abrechnung für die Menschheit" [History, conceived as a pure science and accorded sovereignty, would be for humanity a kind of conclusion to life and a settling of accounts].[11] In its purest form, historical science is only possible after the end of history, or rather, after the end of the history of humanity or the history of life – a similar vision, we can recall here, appears at the high point of Stifter's novel *Indian Summer*. At other points in the essay, Nietzsche mocks the historian's vaunted "objectivity" as the neutrality of the eunuch, but ultimately, the *impotentia* of the historian is the direct correlate of the fact that his object is without the natural limits of life.[12] While the living being has a limited horizon that is "umdunkelt" [darkened] to the past, that past, as

9 Cf. Nadeem J. Z Hussain, "The Role of Life in the Genealogy," in *Nietzsche's On the Genealogy of Morality: A Critical Guide*, ed. by Simon May, (Cambridge: Cambridge UP, 2011), 142–169;; and the section "3.2.1 Power and Life" in R. Lanier Anderson, "Friedrich Nietzsche," in *The Stanford Encyclopedia of Philosophy*, ed. Edward N. Zalta (Stanford: Metaphysics Research Lab, Stanford U, 2017), https://plato.stanford.edu/archives/sum2017/entries/nietzsche/.

10 "Das Unhistorische ist einer umhüllenden Atmosphäre ähnlich, in der sich Leben allein erzeugt, um mit der Vernichtung dieser Atmosphäre wieder zu verschwinden." Ibid. 252 [91].

11 Ibid. 257 [95].

12 Ibid. 284–285 [120–121].

"ein Erkenntnisphänomen" [intellectual phenomenon], remains "todt" [dead] for the one who apprehends it.[13]

By contrast, "monumental" history, for Nietzsche, serves life by providing models for the living. In the manner of *historia magistra vitae*, the greatest achievements of history can teach us what is possible and provide us with the courage to act. Similarly, though in reverse, "critical" history promotes life through self-affirmation. "Critical" historians thus serve their own lives by condemning all historical precedents as a series of crimes and injustices. Both the "monumental" and the "critical" modes of history also pose risks to life, however. One might judge the great monuments of history as the *only* great works possible. Thus historians of art, for example, might suppress contemporary and future greatness in art: "Denn sie wollen nicht, dass das Grosse entsteht: ihr Mittel ist zu sagen 'seht, das Grosse ist schon da!' [...] sie handeln jedenfalls so, als ob ihr Wahlspruch wäre: lasst die Todten die Lebendigen begraben" [For they don't want great art to come into being: their strategy is to say: 'Look, great art already exists!' ... they act as though their motto were 'Let the dead bury the living'].[14] Critical historians also face danger: condemning history does not change the fact that it is our history. Such condemnation only leads to inner conflict: "Wir bringen es im besten Falle zu einem Widerstreite der ererbten, angestammten Natur und unserer Erkenntniss" [we arrive at an antagonism between our inherited, ancestral nature, and our knowledge].[15] Though it is clear that the philosopher has sympathy with this position, he acknowledges that it requires great strength to cut at one's own "roots" and still survive.[16]

If monumentalists admire and imitate the past, while critics curse and condemn it, Nietzsche calls "antiquarian" the "Bewahrenden und Verehrenden" [person who preserves and venerates].[17] Crucially, antiquarians are suited to preserve the past because they have "ein Hindurchfühlen und ein Herausahnen, ein Wittern auf fast verlöschten Spuren, ein instinctives Richtig-Lesen der noch so überschriebenen Vergangenheit, ein rasches Verstehen der Palimpseste, ja Polypseste" [the ability to empathize with things and divine their greater significance, to detect traces that are almost extinguished, to instinctively read correctly a past frequently overwritten, to quickly understand the palimpsests, indeed,

---

13 Ibid. 257 [95].
14 Ibid. 264 [101–102].
15 Ibid. 270 [107].
16 "Dann wird seine Vergangenheit kritisch betrachtet, dann greift man mit dem Messer an seine Wurzeln," ibid.
17 Ibid. 265 [102].

polypsests].[18] Because of this empathy and philological instinct, which encompass all manner of "things" from texts to relics and ruins, including entire places, antiquarian historians help to create a sense of rootedness. Nietzsche explicitly compares "den eigentlich historischen Sinn" [true historical sensibility] to "das Wohlgefühl des Baumes an seinen Wurzeln" [the contentment the tree feels with its roots], and describes how it can protect us from one of the most dangerous aspects of modernity, the "Auswanderungslust" [joy of migration] and the "Suchen nach Neuem und immer Neuem" [craving for new and ever newer things].[19] But, like the monumentalists and the critics, the antiquarian approach has its liabilities. While the antiquarian passion for history protects us from the craven desire for novelty, it can also be stifling. Antiquarians can preserve, but they cannot create new life. The tree dies "wenn der historische Sinn das Leben nicht mehr conservirt, sondern mumisirt" [when the historical sense no longer conserves life but mummifies it].[20]

The metaphor of tree and roots also implies an ecological concept of life, which flourishes only in the appropriate milieu, or under the right atmospheric conditions. Such proto-ecological thought is indeed central to Nietzsche's critique of historicism. His definition of the "ahistorical," for example, is metaphorical: "Das Unhistorische ist einer umhüllenden Atmosphäre ähnlich" [the ahistorical is like an enveloping atmosphere].[21] Just as living beings require an atmosphere to breathe, so too do they require "the ahistorical" to live. Extending the metaphor, Nietzsche also calls this atmosphere a "Dunstwolke" [cloud of mist], which gives the analogy an epistemological turn.[22] It is obvious to the philosopher that the ahistorical, be it in the form of ignorance or false knowledge, generates a "cloud" around an individual or an entire people. Yet while the historically literate see the past more clearly, they also have no air to breathe.

Tree and atmosphere are not the only earthly metaphors in Nietzsche's essay on history. Like the young Stifter, Nietzsche turns to geological metaphors to describe historical change in his own lifetime. Speaking of the "masses," for example, the philosopher suggests that the statistical law of large numbers, for example, only applies to the "lower strata" of society, its "loam and clay."[23] Nietzsche's use of the metaphor of the earthquake, moreover, is key to his overarching argument. Anticipating a central insight of phenomenology that subjects

18 Ibid. 265–266 [103].
19 Ibid. 266–267 [104].
20 Ibid. 268 [105].
21 Ibid. 252 [98].
22 Ibid. 253 [98].
23 "den niederen Lehm- und Thonschichten der Gesellschaft," ibid. 320 [154].

are defined by their "horizons," Nietzsche first establishes that "[die Wissenschaft] alle Horizont-Umschränkungen aufzuheben sucht und den Menschen in ein unendlich-unbegrenztes Lichtwellen-Meer des erkannten Werdens hineinwirft" [science seeks to eliminate all limitations placed on horizons and to catapult the human being into an infinite, unlimited light-wave sea of known becoming].[24] Despite the mixed metaphors, the trend is clear: the solid ground from which one can view the sea as a horizon becomes the high sea, itself dissolved into a field of pure light.

Turning the traditional metaphor of truth as light on its head, Nietzsche thus describes science as a process of liquification and evaporation. In the passage that immediately follows, the philosopher turns to a metaphor that has haunted Enlightenment discourse since the infamous Lisbon earthquake of 1755:

> Wenn er nur darin leben könnte! Wie die Städte bei einem Erdbeben einstürzen und veröden und der Mensch nur zitternd und flüchtig sein Haus auf vulkanischem Grunde aufführt, so bricht das Leben selbst in sich zusammen und wird schwächlich und muthlos, wenn das *Begriffsbeben*, das die Wissenschaft erregt, dem Menschen das Fundament aller seiner Sicherheit und Ruhe, den Glauben an das Beharrliche und Ewige, nimmt.
>
> If only he could live in it! Just as in an earthquake cities collapse and are destroyed and human beings build their houses, but fearfully and fleetingly on volcanic ground, so life caves in on itself and becomes feeble and discouraged when the *concept-quake* unleashed by science robs the human being of the foundation for all his security and tranquility, his belief in what is lasting and eternal.[25]

Nietzsche's metaphorical neologism, "*concept-quake*," echoes the eighteenth century's refunctioning of the term "revolution" to describe geological and political upheavals. The composite term, moreover, underscores the difficulty of dissociating the metaphorical and literal levels of the trope. What exactly quakes in a concept-quake? By analogy to the earthquake, it should be the concept itself, which is as immaterial as the quaking. One thus should ask instead: What scientific knowledge robs the modern human of "his belief in what is lasting and eternal," as Nietzsche writes? As we saw in the introduction, Nietzschean philosophy, at least in the essay "On Truth and Lying the Extra-Moral Sense," written earlier in 1873, reacts to the discovery of geological deep time with a drastic revaluation of human significance. With the concept of the "concept-quake," then, Nietzsche's critique of historicism itself joins the discourse of geohistoricism. The

24 Ibid. 330 [164].
25 Ibid.

logic of earthquakes and volcanos, of rising mountains and eroding valleys, leaves us with a deeper sense of the transience of all being, of all "grounds."

Nietzsche's reception of geological knowledge – in part through Stifter's novel *Indian Summer* – also helps to explain his rather friendly treatment of the antiquarians, despite the philosopher's predictable scorn for "das widrige Schauspiel einer blinden Sammelwuth" [the repugnant spectacle of a blind mania to collect].[26] He accepts the fundamental assumptions behind geohistoricism, namely that geology destabilizes all notions of permanence, but also engenders new practices of preservation and reconstruction. In this sense, the "untimely" thinker is a true contemporary of historicism and thus its most valuable critic. The key concept of this critique, as I have argued, is "life," and in what follows, I would like to ask how the concept of life helps to reveal the limits and internal contradictions of the restorative poetics discussed in this book. To this end, I turn back to Stifter and discuss his attempt, with *Witiko*, to recreate "historical life" in a historical novel.

Witiko: Life and the Historical Novel

Nietzsche's critique of historicism from the perspective of "life" provides a useful framework for evaluating the historicist poetics of his contemporaries, but his concept of life is not easily applicable to a literary work. While the philosopher addresses the question, for example, of how admiration for the monuments of history might stifle creativity as an expression of present life, he does not discuss in the essay on history how works of art and literature serve life aesthetically. If life is tied up with Nietzsche's ethics of self-affirmation, the framework of his critique relies on the distinction between past and present, the key distinction of historicism itself. Since the direct application of a philosophical concept to literary works is also a questionable practice of reading, it is important to consider what notions of life are implicit in Stifter's poetics and literary works, and in particular in his historical novel *Witiko*.

As the conclusion of Stifter's *Indian Summer* demonstrates, life can be the *terminus ad quem* of a literary work. The novel of education leads its protagonist into life, and when it ends with a marriage, as in Stifter's novel, it provides a foundation for future life, both of the protagonist and his descendants: "eines ist gewiß, das reine Familienleben, wie es Risach verlangt, ist gegründet, es wird [...] in ungeminderter Fülle dauern, [...] und jedes selbst das wissenschaft-

26 Ibid. 268 [105].

liche Bestreben hat nun Einfachheit Halt und Bedeutung" [one thing is certain: the pure family life that Risach was referring to has been founded, it will continue undiminished [...] and everything, including my scientific endeavors, has now gained signficantly in clarity, solidity, and importance].[27] In the case of the historical novel *Witiko*, one might say that life operates as a *terminus a quo*, even if the protagonist Witiko, tellingly, also falls in love, marries, and has children. Indeed, as Cornelia Blasberg has written, the character Witiko displays a "vexing similarity" to countless other characters in Stifter's previous work.[28] The purpose of the historical novel, however, is not to recount one person's life, but rather the life of an entire people. In this sense, it is *from* the individual Witiko's life that Stifter aims to represent the larger historical world.

For this reason, the plot of Stifter's historical novel has two topological structures, or rather, two topological approaches to the figure of the circle. The first part of the novel traces the movement from periphery to center. A young man, named simply Witiko, presumably a nobody on the political scene, is sent in 1138 as a messenger to represent the Duke of Bohemia, Slobeslaw, at the quasi-parliamentary Wysehrad in the capital Prague. Because Slobeslaw has fallen ill, the nobility in Prague plan to elect his successor. Confusingly, both pretendants to the position, Slobewslaw's son and nephew, are named Wladislaw. This doubling is historical, but poetically, it underscores the difficulty of the decision, which centers on the competing legal principles of primogenitur (succession of the eldest in direct lineage), and the older tradition of seniority (succession of the eldest in the clan). Ultimately, the latter principle, and hence the nephew Wladislaw, wins the political election in the parliament of higher (*Lechen*) and lower (*Zupanen*) nobility, and approval of his dying uncle, Slobeslaw – but this does not prevent the angered and disinherited son from attempting, after his father's death in 1142, to claim his birthright by military means.

The hero of Stifter's novel, Witiko, supports the official choice, and in the second part of the book fights alongside the reigning (nephew) Wladislaw in the ensuing civil war. But Witiko did not begin as, nor was ever, a partisan in the struggle of succession. Instead, the hero of Stifter's novel is emphatically a messenger. Witiko is always in the right because he always stands on the side of the law (*das Recht*). In his role as messenger, even when he takes up arms, Witiko acts as a medium for the establishment of the political order which has its own topology, an imperial topology of concentric circles. This structure of

27 HKA 4.3, 282 [478–479].

28 Blasberg, *Erschriebene Tradition*, 249. Unlike Blasberg, I do not see this similarity as a form of "repetition" that undermines other meanings of "repetition" in *Witiko*, such as its use as a means to imitate oral poetry.

ever-widening circles motivates both the subplot of the novel, in which Witiko marries, has children, and builds a castle – for which Stifter uses the archaic term *Veste* to underline the symbolic meaning of the castle as a form of political "fortification." Witiko's *Veste* in the Bavarian-Bohemian forest is thus the first circle. The next larger circle is the Duchy of Bohemia, which Witiko defends and helps to establish as a Kingdom. In a third plot line, finally, Witiko fights alongside Emperor Friedrich Barbarossa near Milan, thus defending the rights of the Empire, the third and largest circle. Thus the novel's plot moves not only from periphery to center, but also from family and province to kingdom to empire.

Though the fiction of Witiko's life thus mediates between individual and collective history, it cannot mediate, on its own, between past and present. Yet there are other ways in which life – in both a biological and Nietzschean sense – functions both as the medium and purpose of Stifter's representation of history in the novel. For one, the novel's narratological framework, its prefatory paratext, functionalizes the castle ruins and the living forest as means of time travel, a chronotope that metonymically transports the readers into the past.[29] Second, Stifter's letters, in which he comments on the poetic process of writing *Witiko*, employ anatomical metaphors to conceptualize the now clichéd notion that historical novels "bring history to life." Third, Stifter turns to the literary techniques and visual poetics of realism, its "detail realism" or "reality effects," to accomplish this goal. Here the discourse of geohistoricism leaves its mark on *Witiko*. For the "reality" of history resides in its material traces and must be reconstructed geologically and archeologically. Fourth and finally, however, the competing demands of historicist realism create problems as well, which are best understood as the conflict between philology and literary history. The literary techniques that "enliven" history are themselves not historical, whereas the "authentic" historical sources are not alive.

Framing History: The Castle of Wittinghausen

A castle ruin in the midst of a vibrant forest is a familiar topos in Stifter's literary works. One such castle in a forest appears in the novella *Castle Crazy* (*Die Nar-*

29 Although Mikhail Bakhtin relates his concept of *chronotope* to Einsteinian "space-time," geology presented an earlier and arguably more culturally influential form of "time in space." Cf. "Forms of Time and of the Chronotope in the Novel," in *The Dialogic Imagination. Four Essays*, ed. Michael Holquist, trans. Caryl Emerson and Michael Holquist (Austin: U Texas P, 1981), 84–85.

renburg), which I discussed in the first chapter as an early example of Stifter's geohistoricism. A key aspect of *Castle Crazy* was the intertwining of architectural restoration and genealogical repair. The heir of the "crazy castle" is capable of historical reconstruction, but he is also, crucially, a legal and biological heir. As we just saw, the plot of *Witiko* also centers on questions of genealogy, of biological and legal succession. As in the novella *Castle Crazy*, blind obedience of the law of primogenitur functions both as an ideal and a source of trouble in *Witiko*. In the historical novel, the principle of primogeniture first causes a war of succession, then becomes codified in the new hereditary monarchy. Although the civil war in *Witiko* serves as a warning against the dangers of competing genealogical claims, the novel on the whole adopts a reverent attitude towards the preeminence of genealogical ties in medieval feudal culture. Indeed, by depicting the transition from an "electoral" to a "hereditary monarchy," Stifter's novel represents a conservative reaction to the emergence of constitutional monarchies and republicanism in the nineteenth century.[30] On a textual level, the novel defends genealogical order through a (metaphorical) semantics of "roots" or "line" (*Stamm*). The *Stamm* of the Premysl, and Witiko's reinvigoration of his own, minor *Stamm* of the Rosenbergs, are of primary importance.

In *Witiko* Stifter forgoes the fictional framing narrative of a present day heir, unlike in *Castle Crazy*, or his other major work, the novel *My Great-Grandfather's Notebook*. *Witiko* thus also does away with the editorial fiction of found manuscripts present in those works.[31] Instead, as the short preface to the historical novel shows, Stifter's relationship to the story of Witiko is geohistorical: "In meiner Kindheit traten mir schon öfter Spuren eines Geschlechtes entgegen, das im mit täglichen Böhmen gehaust hat, und in der Erinnerung und in den Er-

30 Albrecht Koschorke, "Bewahren und Überschreiben. Zu Adalbert Stifters Roman *Witiko*," in *Vergessene Texte*, ed. Aleida Assmann and Michael C. Frank (Constance: Universitätsverlag Konstanz, 2004), 148. Similarly, Horst Albert Glaser, "Auf Witikos Spuren. Stifters Schlachtenschilderungen," *Germanistische Mitteilungen* 40 (1994): 186, claims that *Witiko*, in contrast to Goethe's *Götz*, legitimatizes centralized monarchal rule over earlier forms of relative territorial independence. Such views correct earlier research that argued Stifter's intention was to depict a "utopia" of political dialogue, cf. Ferdinand Seibt, "Stifters 'Witiko' als konservative Utopie," in *Deutsche und Tschechen. Beiträge zu Fragen der Nachbarschaft zweier Nationen.*, ed. Adalbert-Stifter-Verein Munich (Munich: Delp, 1971), 23–29; and Martin Selge, "Die Utopie im Geschichtsroman. Wie man Adalbert Stifters 'Witiko' lesen kann," *Der Deutschunterricht* 27, no. 3 (1975): 70–86.

31 As a result, the novel simulates historical speech as oral speech, cf. Kai Kauffmann, "Mündliche Rede, schriftliche Urkunde, wörtliches Protokoll. Adalbert Stifters Geschichtsroman *Witiko.*," in *Grenzräume der Schrift*, ed. Achim Geisenhanslüke and Georg Mein (Bielefeld: transcript, 2008), 128, and my discussion below.

zählungen des Volkes fortlebte" [In my childhood, traces of a clan often appeared to me, that had its home in southern Bohemia, and lived on in the memory and in the stories of the people]. At issue here is not the continuity of a family in its genealogy, but as a "trace" (*Spuren*) and as a subject of oral history and literature. Stifter's poetic restoration of these "traces" begins early: "Als Jüngling ging ich diesen Spuren nach, und habe manchen Tag in den Trümmern der Stammburg dieses Geschlechtes zugebracht" [As a youth, I followed these traces and spent many days among the ruins of the main castle of this clan].[32]

Most interesting in this description of the poetics of geohistoricism is the next sentence, which ascribes poetic agency to the ruins: "Hierauf strebte das Ding sich in verschiedenen kindischen Versuchen dichterisch zu gestalten" [On these occasions, the thing strove towards poetic form in several immature attempts].[33] "Thing" or *Ding*, arguably Stifter's favorite word, refers here not only to something undefined and perhaps undefinable, but also specifically to the material remains of the past.[34] As bare material, traces of rock and in rock are, at least at first glance, merely "things." In Stifter's restorative poetics, the energy, the striving (*streben*), comes from this thing that seeks linguistic and poetic form. The agency of the ruins recalls other passages in Stifter's work: the antiquities that would speak more eloquently in their original environment than at the "Flea Market;" and the history of the earth, which it carries in its own "room of records." Indeed, one can even think of the "Dichtung des Plunders" [poetry of rubbish], and interpret it as a *genitivus subjectivus:* the rubbish has its own poetry to tell.[35] It is as if the story of Witiko is there, in the ruins of Wittinghausen; it just needs to "take form," or as Stifter writes: "Später fand sich, begleitet von mancher Unterbrechung und Wiederaufnahme, etwas Ernsteres zusammen, und ging in jüngster Zeit der Vollendung entgegen" [Later, after taking it up with repeated interruptions it took a more serious form; a short while ago it was nearing completetion].[36]

Readers of Stifter's work will also recall that the castle of Wittinghausen has already served as an inspiration for his historical novella *The High Forest* (*Der*

32 HKG 5.1, 11 [7].

33 Ibid. Translation modified. Frye renders the sentence in the active voice: "On these occasions, I made several immature attempts to give poetic form to their story."

34 Blasberg notes here, "*Witiko* ist auch Ruinenpoesie," *Erschriebene Tradition*, 283. If I largely agree with her intertextual reading of *Witiko* as a "novel of literary history," the geohistorical approach places the "too" at a different position. The *traditum* of *Witiko* is material (ruins), and it is literary, *too.*

35 HKG 1.5, 16.

36 HKG 5.1, 11 [7].

Hochwald). The novella may be one of the works that Stifter considers "childish." In a letter to Heckenast, Stifter admits that "as a frivolous young man" he "broke history over his knee, and grafted it into the desk drawers of my fantasy."[37] As we shall see, the metaphor of "grafting" is one of many organic metaphors employed by Stifter to distinguish *Witiko* from his other works. If the motif of the forest is central to both the youthful novella and the mature historical novel, it is because the latter avoids the earlier work's artificial graft and desk-bound fantasy.[38] The name Witiko means "forest dweller," and, we are repeatedly told, he leads a "forest people."[39] Witiko and his people live in and from the forest, derive their strength from the forest, and even "denken wie der Wald" [think like the forest].[40]

Before we learn any of all this, however, the narrator of *Witiko* must first lead us into the forest. Stifter already employs this technique of heterodiegetic description of setting in the *High Forest.* Though a common technique of literary realism, it is nonetheless remarkable in these instances as a geohistorical technique. In both stories, the distance of the narrator from the narrated events is primarily temporal. Consider the first line of *Witiko:* "Am oberen Laufe der Donau liegt die Stadt Passau" [The City of Passau is situated on the upper part of the Danube].[41] The present tense marks the time of the narrator, who concludes the introductory description of setting two pages later with another time marker ("Zur Zeit") and a switch to the preterit: "Zur Zeit, da in Deutschland der dritte Conrad, der erste aus dem Geschlechte der Hohenstaufen, herrschte" [at the time when Conrad III, the first of the Hohenstaufen clan, ruled in Germany].[42]

---

37 Stifter to Heckenast, March 7, 1860, "Die Geschichte als leichtsinniger Mensch über das Knie gebrochen, und sie dann in die Schubfächer meiner Fantasie hinein gepfropft," PRA 19, 223–224.

38 Cornelia Zumbusch thus reads the novel's "genetic narration" as an attempt to "naturalize" history, "Der Gang der Geschichte: Historismus und genetisches Erzählen in Stifters 'Witiko,'" *Jahrbuch der deutschen Schillergesellschaft* 51 (2007): 228. On "genetic narration," cf. Jörn Rüsen, "Die vier Typen historischen Erzählens," in *Formen der Geschichtsschreibung*, ed. Reinhart Koselleck (Munich: dtv, 1982), 514–605.

39 Cf. Klara Schubenz, "Botanik/Wald," in *Stifter-Handbuch*, ed. Begemann and Guiriato, 260–261; and Schubenz: *Der Wald in der Literatur des 19. Jahrhunderts: Geschichte einer romantisch-realistischen Ressource* (Göttigen: Wallstein/Konstanz UP, 2020), chapter 6.2.

40 HKG 5.1, 57, translation modified from: "I do know the ways of the forest," [41]. Cf. Birgit Ehlbeck, *Denken wie der Wald: Zur poetologischen Funktionalisierung des Empirismus in den Romanen und Erzählungen Adalbert Stifters und Wilhelm Raabes* (Bodenheim: Philo, 1998), 156–175.

41 HKG 5.1, 13 [13].

42 HKG 5.1, 15. Frye translates this subordinate clause with "During the reign of the German King, Konrad III," [14].

Of course, the city of Passau was also situated on the Donau when Conrad ruled over the Holy Roman Empire in the twelfth century. The course of the rivers Donau and Moldau, which lead the narration into the Bohemian-Bavarian forest, has also hardly changed. What about the forest itself? This question is at the heart, I would argue, of how *Witiko* establishes itself as a historical novel. In *The High Forest*, it is important that the forest, and specifically the "high forest," has not changed. The novella recounts an attack on the Wittinghausen castle by Swedish troops during the Thirty Years' War, which the two daughters of the house escape by fleeing into the "high forest," only to return to find their father and brother killed. The events of the novella thus explain how Wittinghausen became a ruin, which as a motif, represents a counterpoint to the motif of the forest. In the novella's most dramatic scene, the two girls watch, from the safety of the woods and through the medium of telescope, as their home is burned down.

The "high forest" is thus immune to the historical change that befalls the castle. It is also a symbol of untouched or "virgin" nature, which protects the virgin girls, who, in a plot twist with an overdetermined meaning, never marry and live out their lives as spinsters in the barren castle ruins. Yet this untouched and unchanging natural space also has mediating power: Stifter describes the lake where the girls take refuge metaphorically as a *Naturauge* [eye of nature]. Combined with the telescope, the metaphor of an optical medium suggests that the forest can mediate between the present observer and past event and deliver, moreover, a much clearer image than the ruin itself. The ruin is a sign of the past, but an obscure and indirect one: when the girls "see" it burning, they actually only see a cloud of smoke, a sign to be interpreted. And where there are smoke and fire, there will also be an investigation. Reconstructed from the ashes of Wittinghausen, the story of the *High Forest* is comparable to the report of a fire marshal or the studies of archeologists at Pompeii.

In *Witiko*, ruin and forest also play a key role in the mediation of past and present. At first glance, the narrative frame appears to be the same as in *The High Forest*. In the preface, Stifter declares that he was inspired by the ruin to tell its story – in this case, the story of how the castle was first built, that is, the beginning of the story whose end he told in *The High Forest*.[43] In the novel's introductory paragraphs, the narrator establishes geographic continuity between the present and the past. When Stifter describes the forest, however, he emphasizes how much it has changed: "obwohl es jetzt zum größten Theile mit Wiesen, Feldern und Wohnungen bedeckt ist, so gehört es doch dem Hauptwalde an, mit dem es

43 Consequentially, Stifter titles a chapter in the third volume of *Witiko* "Im hohen Walde" [In the High Forest].

vielleicht vor Jahren ununterbrochen überkleidet gewesen war" [Even though it is now covered primarily by meadows, fields, and farm buildings, it may still be reckoned to the forest proper since many years ago it was more than likely completely covered by trees].[44] Other sentences in the novel's introduction rely on the self-evident knowledge that historical progress causes deforestation: "Dieser auch jetzt noch große Wald" [Even today the forest is quite extensive].[45]

Such proto-ecological thought recalls Stifter's early cultural criticism in *Vienna and the Viennese*, though it resonates most strongly with his late novella *Nachkommenschaften* [*The Descendents*, 1864], which portrays the drying out and destruction of a marsh landscape by a family aptly named Roderer, or, the "ones who clear (forest) land" (*roden*). In *Witiko*, moreover, this thought does not remain on the level of the authorial-narrative voice. It also comes from the mouth of Huldrik, an old servant of Witiko's family. Like his name suggests (*huldigen:* "to praise and honor"), Huldrik's extreme, messianic devotion to his masters recalls the old servant from *Castle Crazy*. And as in the earlier novella, this old man tells the young Witiko about the deep past of his family's genealogy, which extends back into "the olden times" when the pagan gods Thor and Freia still reigned.[46] So old is the tale of these ancient times that it is not passed down from father to son, but from great-grandfather to great-grandson. It is a story of thousands, not hundreds, of years. In this deep prehistory, there was "gar kein Wald" [no forest at all] in Witiko's homeland. Only later, after "kriegerische Männer aus Welschland" [warlike men from Italy] drove out the Germanic peoples and destroyed everything, does the forest return: "es ist der Wald gewachsen, als wäre nie etwas anderes da gewesen" [the forest grew as if there had never been anything else].[47]

Huldrik's prophetic history of the Witigon clan establishes an inverse relationship between the success of the clan and the thriving of the forest. On a metaphorical level, however, there is a direct relationship between the success of a family and the thriving of a "root" or *Stamm*. Even though the forest people literally destroy the forest, metaphorically, *they are the forest*. Thus Huldrik calls the first Witiko, who brought Christianity back home from Rome, a *Sprößling*, a sprout or offshoot, and employs the word *Stamm* to refer to the "clan" (the only possible one-word translation for the lexicaled metaphor): "es ist geweissagt worden, daß immer ein Witiko den Stamm erretten werde" [It was prophe-

44 HKG 5.1, 14 [13].
45 Ibid.
46 "In alten Zeiten, als noch Thor und Freia herrschten," HKG 5.1, 228 [148].
47 HKG 5.1, 228–229 [148].

sized that a Witiko will always save the clan].[48] The metaphor of the genealogical "tree" thus suggests continuity across history. This link can only be broken if the stem or root itself dies out – a danger that Stifter, himself childless, thematizes obsessively in his novellas, from *The Bachelor* to *The Descendants*. The metaphor of the "forest," however, and in particular the "main forest" or "high forest," appears to guarantee that the extinction of the Witigons is unlikely, if not impossible. The forest represents the cyclicity of nature and transfers this cyclicity to its inhabitants. Indeed, Huldrik ecstatically confirms the fulfillment of his prophecy in the third part of the novel: "Witiko, Witiko, ein Witiko wird kommen, und Witiko ist gekommen" [Witiko, Witiko, a Witiko will come, and Witiko has come].[49]

How can these two views of the relationship between forest and forest people, between nature and culture, be reconciled? One possibility is to deemphasize the importance of proto-ecological thinking in Stifter's work, or at least in *Witiko*. As Klara Schubenz argues, the protagonist Witiko sees the forest as an economic "resource" to be exploited.[50] At the same time, the tone of the novel's introduction communicates nostalgic regret at the loss of the forest, a regret similar to that expressed in *The Descendants*. The semantics of the term *groß*, like the English "great," signals this regret because it suggests both size and importance. Moral, cultural, and political greatness are central themes of *Witiko*, which, I will argue in further detail below, is best understood as a prose epic, a genre that depicts greatness in a grand style. In the introduction to *Witiko*, "greatness" is not only a quality of the forest, or more precisely, a quality that once described the forest, but also of the towns within it, which are marked by a form of decline that has personal meaning for Stifter: "Die bedeutendsten Orte, denen sie [die Moldau] in dem Laufe, der genannt worden ist, in den heutigen Tagen begegnet, sind die Flecken Oberplan und Friedberg, die Abtei Hohenfurt und die Städte Rosenberg und Krumau" [Nowadays, the most important places you would encounter along this course are the villages (*Flecken*) of Oberplan and Friedberg, Hohenfurt Monastery, and the cities of Rosenberg and Krumau].[51] The superlative "most important" is relative, depending on both the restriction of place and time. As the author's birthplace, Oberplan is certainly a "most important" place "nowadays." By fondly calling the village a "spot" (*Fleck*), however, Stifter acknowledges Oberplan is not an "important place" in the context of nineteenth-century geopolitics. The story of Witiko is thus de-

48 HKG 5.1, 229 [148].
49 HKG 5.3, 97 [451].
50 Schubenz, "Botanik/Wald," 261.
51 HKG 5.1, 15 [14].

signed to demonstrate that in the twelfth century, the "important" towns in the Bavarian-Bohemian forest were also significant on a grander political scale. And if this demonstration succeeds, the author of *Witiko* will bring fame back to his hometown in the present, as well.

Read in this way, the central political meaning of Stifter's historical novel is an expression of local patriotism – an intention the author establishes in the book's dedication. It is reasonable to suspect, then, that its central poetic meaning relates to the narrative of decline and loss. Paradoxically, as mentioned before, the metaphors that associate the forest with Witiko's family establish an inverse relationship between their respective growth: when the forest people thrive, they exploit and destroy the forest. Moreover, when Witiko's people are driven away by the Romans, the forest regrows and it appears "as if there had never been anything else" there. The forest thus also obscures the "traces" of a family that are so dear to the author and his childhood fantasy of reviving this history. This presents a dilemma: Should Stifter "restore," through his writing, Castle Wittinghausen or the forest that once surrounded it? In restoring local history to its former greatness, must the offspring of Oberplan also acknowledge the history of its decline? In its introductory descriptions of the Bavarian-Bohemian forest, *Witiko* appears to rely on milieu or "atmospheric realism" in order to transport its nineteenth-century readers into the twelfth-century past. By name, and by his character of quiet strength, the hero Witiko harmonizes with his forest environment, as well. But Witiko is not like the other "forest-dwellers" in the novellas set in Stifter's present day: a figure modeled on the heroes of medieval German epics can be neither a naturalist as in *Castle Crazy* or the *Kiss of Sentze*, nor a landscape painter as in *The Descendants*, nor an austere hermit as in *Abdias* or *Limestone*. Witiko may decorate his shield with a "forest rose," but he does not fight on its behalf.

"Among Clubs and Axes:" The Birth of a Novel in the Museum of Cultural History

If the forest cannot mediate between present and past – or at least not in a simple way – where do things stand with the ruin? Through the analysis of the discourses of art historical restoration as they are expressed in Stifter's *Indian Summer* and Viollet-le-Duc's theoretical writing, I have attempted to show how geohistoricism comes to view history as a collection of objects. Assembled into a coherent milieu, restored to their original places, the things of the past begin to "speak" to us. The flipside of this vision becomes apparent in *Witiko:*

instead of a convincing description of milieu, many readers have complained that the novel amounts to a catalog list of historical objects.[52]

Long, detailed lists of people, places, and objects are characteristic of epic poetry. As such, the catalog is one of the features of *Witiko* that defy the poetic expectations of the modern novel for the sake of reviving the epic tradition – and hence a source of frustration for modern readers.[53] The most insightful analysis of the relationship between nineteenth-century realism and material objects remains Roland Barthes's classic discussion of the "reality effect." Barthes's argument revolves around the distinction between the "reality effect," and what he calls the "referential illusion."[54] The mere naming of objects in a literary text suffices to create a referential illusion, which, as a result, is a feature of all genres of fiction, from epic to fable, novel and fairy tale. Yet this universality is also the reason that Barthes calls the reference function of fiction an illusion: we know that the word "table" is not the same as a table. The production of the "reality effect," through which a text signifies not individual things but the category of the real, requires a special structure. Barthes cites the example of the barometer on the table of Madame Aubain in Flaubert's tale *Un cœur simple* [*A Simple Heart*, 1877]. Unlike the other signs in the text, Barthes argues, the barometer is empty of meaning. It does not say anything particular about the person who owns it, nor about the place where it stands. If it had such meaning, it would be part of the referential illusion of the realist novel.

We can recall here Balzac's programmatic discussion of "milieu"-realism. Balzac discusses the concept of milieu[55] in his *avant propos* to the 1842 edition of the *Comédie Humaine* [*The Human Comedy*]. Of his inspiration for the *Human Comedy*, Balzac writes: "Cette idée vint d'une comparaison entre l'Humanité et l'Animalité" [The idea occurred to me, in consequence of a comparison between

52 Helena Ragg-Kirkby, who calls *Witiko* a "manic masterpiece," is one of the few who elevates the frustration to a positive aesthetic value of Stifter's proto-modernism, *Adalbert Stifter's Late Prose*, 78.

53 Alternatively, one can see in the paratactic and metonymical, and thus seemingly "contingent" and "arbitrary" linking of signs a first step towards the "lexeme-autonomy" of modern literature, cf. Moritz Baßler et al., *Historismus und literarische Moderne* (Tübingen: Max Niemeyer, 1996) 58.

54 Roland Barthes, "The Reality Effect," in *The Rustle of Language* (Berkeley: U California P, 1989), 148.

55 On the historical semantics of the word *milieu* see Leo Spitzer, "Milieu and Ambiance: An Essay in Historical Semantics," *Philosophy and Phenomenological Research* 3, no. 1 (September 1942): 1–42.

the human and animal kingdoms].[56] In this case, Balzac borrows his concept of zoology from Geoffroy Saint-Hilaire, for whom the *milieu ambiant* plays an essential role in the differentiation of organic forms. Although Geoffroy Saint-Hilaire and Cuvier famously butted heads in an 1830 debate over the theory of evolution, the role of the *milieu* as a determining fact in morphology fits into both arguments. As Balzac summarizes Geoffroy Saint-Hilaire's theory: "L'animal est un principe qui prend sa forme extérieure, ou, pour parler plus exactement, les différences de sa forme, dans les milieux où il est appelé à se développer. Les Espèces Zoologiques résultent de ses différences" [It is the principle of animal life that each creature should borrow its external form, or, to speak more exactly, the differences of its form, from the environment (*dans les milieux*) in which it is destined to become developed. The zoological species result from these differences].[57] The *Human Comedy* is thus based on this principle, applied to social life and social milieus: "La Société ne fait-elle pas de l'homme, suivant les milieux où son action se déploie, autant d'hommes différents qu'il y a variétés en zoologie ?" [Does not society make of man, in accordance with the environment (*suivant les milieux*) in which he lives and moves, as many different kinds of men as there are different zoological varieties?][58]

Balzac's concept of milieu is not only biological and social, but also historical (likely a reason for his preference here for Geoffroy Saint-Hilaire, who was more open to evolutionary ideas). As an admirer of Walter Scott's historical fictions, Balzac envisions the *Human Comedy* as a vast historical novel, a history of his own time: "Avec beaucoup de patience et de courage, je réaliserais, sur la France au dix-neuvième siècle, ce livre que nous regrettons tous, que Rome, Athènes, Tyr, Memphis, la Perse, l'Inde ne nous ont malheureusement pas laissé sur leurs civilisations" [Thus, by patience and perseverance, I might be able to realize for France in the nineteenth century, the book we so much regret, that Rome, Athens, Tyre, Memphis, Persia and India have not left us to tell us of their civilizations].[59]

In his famous study of realist poetics, *Mimesis,* Erich Auerbach ascribes this connection between realism and the historical novel to the role played by the milieu or "atmosphere" in each genre: "Atmosphärischer Historismus und atmosphärischer Realismus hängen eng zusammen" [Atmospheric historicism and at-

56 Balzac, "Avant Propos," (1842) *La Comédie Humaine,* 7; Balzac, *Scenes of Parisian Life,* vol. 1, trans. Ellery Sedgwick (Philadelphia: George Barrie & Son, 1896), iii.

57 Balzac, "Avant-propos," 8; "Author's Preface," iv.

58 Ibid.

59 Ibid. 11 [x].

mospheric realism are closely related].[60] Auerbach's analysis of Balzac centers on the novel *Le Père Goriot*, and in particular the inn or *pension* of Madame Vauquer. Balzac describes Mme. Vauquer and the inn she owns such that "toute sa personne explique la pension, comme la pension implique sa personne" [the person explains the inn, just as the inn implies the person].[61] As Auerbach writes, Balzac's poetics fuses the character and the place into a harmonious whole, "eine atmosphärische Stileinheit" [an atmospheric unity of style].[62] This "unity of style" epitomizes the two foundations of modern realism as Auerbach defines it: the representation of everyday reality and people, and "die Einbettung der beliebig alltäglichen Personen und Ereignisse in den Gesamtverlauf der zeitgenössischen Geschichte, der geschichtlich bewegte Hintergrund [the integration of random, everyday people and events into the process of contemporary history as a whole, the historically dynamic background].[63] Through this interweaving of foreground and background, nineteenth-century French realism – above all Stendhal, Balzac, and Flaubert – thus closes the arc of Auerbach's larger argument. Modern realism fulfills the promise of two ancient styles, the Homeric with its "Vordergründlichkeit" (foregroundedness) and the Biblical with its "Hintergründlichkeit" (backgroundedness).[64] The concept of milieu – *milieu*, the mid-place or the place in-between – thus describes a dialectic of foreground and background at work in modern realism as a kind of *Zwischengründlichkeit* or middle-groundedness, so to speak.

In the context of milieu realism, all signs contribute to the characterization of a milieu. To return to Roland Barthes's formulation, the meaningless detail is an object which is neither significant in its presence nor absence, except that it creates a reality effect. Such details do not characterize the milieu, but rather signify that it is real. In the historical novel, however, or in Balzac's realism, which understands itself as a historical novel of the present day, this "meaningless" detail has an additional message to convey. Such details say, "We are the real," but they also say, "We are the historical." In Barthes's terminology, the realist historical novel has "the real" as its "signified of connotation" but retains, by designating a historical milieu, a global "signified of denotation" or "referential illu-

60 Erich Auerbach, *Mimesis: Dargestellte Wirklichkeit in der abendländischen Literatur*, 10th ed. (Tübingen: Francke, 2001), 441.

61 Balzac, *Le Père Goriot*, quoted in Auerbach, *Mimesis*, 437.

62 Auerbach, *Mimesis*, 441.

63 Ibid. 458.

64 Ibid. 26–27.

sion."[65] These "historical effects" are so commonplace in the historical novel that their overabundance tends to cheapen or trivialize the genre. Any story can be "historicized" so long as one includes reference to the proper clothing, technology, architecture, cuisine, dialect, etc. of a past time and place.

Stifter bemoaned such clichéd practices that turn history into a mere backdrop: "Man erzählt gewöhnlich bei geschichtlichem Hintergrunde Gefahren Abenteuer und Liebesweh eines Menschen oder einiger Menschen. Mir ist das nie recht zu Sinne gegangen" [Usually, history is the background to stories of danger adventure and heartbreak of one person or a few people. I never quite saw the point].[66] "Adventure" and "heartbreak" are the stuff of popular or low-brow historical novels, which Stifter clearly wished to avoid. The reference to "one person or a few people" is also significant, however. Stifter favored Walter Scott's novel *Old Mortality* (the second of the *Tales of My Landlord*, translated into German as *Die Presbyterianer*) because of its focus on the *collective* fate of a people or "das Völkerleben in breiteren Massen" [the life of peoples in greater masses].[67] Following a model of historical thought inaugurated in the eighteenteenth century by Johann Gottfried Herder, Stifter imagines "Völker als großartige Naturprodukte" [peoples as magnificent products of nature], whose development follows a moral law analogous to the natural laws that govern earth history: "in ihren Schiksalen zeigt sich die Abwiklung eines riesigen Gesezes auf, das wir in Bezug auf uns das Sittengesez nennen, und die Umwälzungen des Völkerlebens sind Verklärungen dieses Gesezes" [in their destinies the unrolling of a great law becomes manifest, which we call, as it applies to us, moral law, and the revolutions of the life of peoples are the transfigurations of this law].[68] This appeal to a "great law" recalls Stifter's elaboration of a "gentle law" in his preface to the *Multi-Colored Stones*. In both cases, the analogy between natural and moral laws places Stifter not only in the tradition of the philosophy of history from Herder to Hegel, but also well within the discourse of

---

65 Barthes, "The Reality Effect," 148. For this reason, Barthes associates the "referential illusion" with the poetics of the historian Adolphe Thiers, ibid., see also Barthes, "The Discourse of History" in *The Rustle of Language*, 139.

66 My translation. PRA 19, 282.

67 "Mir haben unter Walter Scotts Romanen die am besten gefallen, in denen das Völkerleben in breiteren Massen auftritt wie z. B. in den 'Presbiterianern.'" Ibid. In this assessement, Stifter agrees with contemporary literary critics such as Wolfgang Menzel and Julian Schmidt, cf. Alfred Doppler and Wolfgang Wiesmüller's commentary, "Historischer Roman und Epos," in: HKG 5.4, 247.

68 PRA 19, 282.

mid-nineteenth-century historicism and its "genetic principle."[69] Without formulating an explicity theory of the milieu, as Balzac does, Stifter likewise grounds his understading of the "life of a people" in the biological milieu: Witiko's people are forest people.

The Body of the Middle Ages

As the complexity of the forest motif in *Witiko* demonstrates, however, genetic and even genealogical models of history do not translate seemlessly into a principle of narration. Stifter certainly approaches the writing of *Witiko* with scientific rigor: "Der Roman hat eine wissenschaftliche Seite, die von vorn herein in keines Menschen Seele liegt, sondern die er sich erwerben muß, *das Geschichtliche*" [The novel has a scientific side, which the writer does not hold in his soul, but rather must acquire: *the historical*].[70] Yet Stifter's historical research borrows elements from several approaches to the past, which are neither interchangeable nor entirely compatible. In researching the events of the novel, he relied heavily on accounts by professional historians, above all Franz Palacky's *History of Bohemia*.[71] Yet Stifter also consulted, or at least claimed to consult, older sources, including primary sources, which confronted him with philological problems of translation and modernization.[72] As Cornelia Blasberg has shown, "the historical" of *Witiko* also encompasses an idealized literary history, which motivates Stifter not only to feature the writing of the *Nibelungenlied* in his plot, but also to model his novel on the genre of the epic.[73]

69 Cf. Linda Simonis, *Genetisches Prinzip: Zur Struktur der Kulturgeschichte bei Jacob Burckhardt, Georg Lukács, Ernst Robert Curtius und Walter Benjamin* (Tübingen: Max Niemeyer, 1998).
70 PRA 18, 169.
71 Franz Palacky, *Geschichte von Böhmen. Größtenteils nach Urkunden und Handschriften. Erster Band. Die Urgeschichte und die Zeit der Herzöge in Böhmen bis zum Jahre 1197* (Prague 1836). Since Karl Flöring, *Die historischen Elemente in Adalbert Stifters "Witiko"* (Gießen: Münchow, 1922), research on Stifter's sources has amply documented the passages in *Witiko* that borrow from this and other historical works. The commentary in HKG 5.4 provides a detailed summary, 226–241.
72 Flöring notes Stifter's tendency to borrow only from original sources when they appear in footnotes, *Die historischen Elemente*, 20. Linguistic barriers clearly played a role. Commenting on Ottokar von Horneck's *Reimchronik*, which he consulted in an 1821 anthology edition by Theodor Schacht, Stifter bemoans how he "[...] oft tagelang in der widerstrebendsten Sprache lesen muß (in einer oft verzweifelten Weitschweifigkeit) um nur ein paar Züge für sich zu erhaschen" [must often read for days in the most averse language (of despairing prolixity) in order to snatch a few passages for himself.] PRA 17, 303.
73 Cf. Blasberg, *Erschriebene Tradition*, see below.

In order to research and transport the "reality" of the twelfth century, however, Stifter turns to cultural history or *Culturgeschichte*. Spanning the whole spectrum of the everyday, cultural history vastly expanded the purview of the historical sciences. As Jacob Burckhardt writes, "the sudden significance of countless individual data points" constitutes a seemingly "endless field" of study.[74] "Cultural history" investigates everything that was not included in what was previously the exclusive domain of history or *Weltgeschichte*, namely state, political and military history, as well as the lives of regents. Indeed, for Burckhardt, cultural history comprises the largest part of "the historical" or "geschichtliche Leben, wie es tausendgestaltig [...] daherwogt" [historical life, as it rolls in a wave of myriad forms].[75] While there is a specific philosophical tradition of cultural history that has its roots in the works of Vico and Herder, the rise of historical geology and archeology in the nineteenth century shifted the focus of cultural-historical research towards material objects and everyday life. And despite the appeal of genetic models that provide an overarching theory of cultural change, they do not, or only partially, explain the striking differences between the material objects of past and present.

As material objects played an increasingly important role in cultural history, they also led to the founding of a new type of museum collection. In 1833, Anton von Spaun, who would later become an admired friend of Stifter, helped found the *Verein des vaterländischen Museums für Österreich ob der Enns* [*Association of the National Museum for Austria above the Enns*], which eventually became the Landesmuseum [State Museum] Linz.[76] As early as 1848, long before focusing his novel project on the character Witiko, Stifter wrote to his editor Heckenast: "Ich arbeite schon an 'Rosenberg' im hiesigen [Linzer] Musäum size ich wöchentlich 2 Mal unter Kolben Äxten Schwertern ... und studire in Horneks Reimchronik" [I am already working on 'Rosenberg' in the museum here [in Linz] I sit twice a week among the clubs, axes, swords... and study Horneck's *Rhymed Chronicle*].[77] This scene is emblematic of the remarkable blend of historical discourses that will result in the novel *Witiko*. In the first instance, history speaks to Stifter through material things, the clubs, axes, and swords hanging in the museum. Second, historical knowledge still resides in a text – in this case the fourteenth-century *Styrian Rhymed Chronicle* (*Steirische Reimchronik*) by Ottakar (*Otacher oûz der Geul*), a historical chronicle in verse, which, in addition to po-

74 Burckhardt, *Über das Studium der Geschichte*, 84. See Chapter 1.

75 Jacob Burckhardt, "Über geschichtliches Studium," in *Gesammelte Werke*, vol. IV (Basel: Schwabe, 1956), quoted in Zumbusch, "Der Gang der Geschichte," 232.

76 E. Lebensaft, "Spaun, Anton von (1790–1849)."

77 PRA 17, 286.

litical history, contains important descriptions of medieval court life.[78] These divergent and at times competing discourses raise questions about the media that best communicate history: Which mediation "preserves" history best? Which offers the greatest imaginative access?

Stifter's museum visits shaped the final text. The novel's descriptive passages at times resemble museum inventories. Like a museum conservator, moreover, Stifter rigorously maintains the narrative mode of external focalization: the narrator only describes what he can see.[79] Here is a passage from one of the earliest social scenes in the novel, as rider Witiko arrives at an inn:

> Auf der Bank, die vor dem Hause hinlief, saß ein Mann, von dem Halse bis zur Sohle in das gleiche Stück groben braunen Tuches gekleidet. Das Tuch lag fest an seiner schlanken Gestalt. Um die Schultern hatte er ein sehr kürzes Mäntelchen mit Ärmeln, das von grauer Farbe war, und noch gröberes Tuch zeigte als die andere Bekleidung. Schwere Schuhe hüllten die Füße ein. Sonst hatte er nichts auf seinem Körper. [...] Der Mann band mit seinen Händen einen festen Eisendraht gitterartig um einen geklüfteten irdenen Topf. Der Reiter saß mit seinem Angesichte dem Manne gegenüber.
>
> Seitwärts des Ritters, etwa zehn Schritte von ihm entfernt, saßen an dem Brettertische zwei andere Männer. Sie hatten sehr beschmutzte Lederkoller an. Die untere Bekleidung konnte man der sehr breiten Tischplatte willen nicht sehen. Ihre Lederhauben lagen auf dem Tische. [...] Vor ihnen stand ein großer grauer Steinkrug mit blauen Blumen. An der Bank neben dem Tische lehnte eine Armbrust, auf der Bank aber lag ein eisenspitziger Stock, den man auch einen Speer nennen konnte.

> On the bench in front of the house sat a man clad from head to toe in a simple piece of coarse brown cloth. The garment fitted snugly on his slender figure. Around his shoulders he had a very short gray jacket with sleeves, made of coarser material than his tunic. He had heavy shoes on his feet. He was wearing nothing else. [...] The man was winding a substantial iron wire-like grating around a split earthen jug. The horseman [Witiko] sat down facing the man.
>
> About ten paces to the side of the horseman two other men were sitting at a wooden table. They were wearing very dirty leather doublets. Because of the wide table, what they were wearing below the waist wasn't visible. Their leather helmets were on the table [...] A large gray mug painted with blue flowers was on the table in front of them. A crossbow was leaning against their bench; on the bench itself was an iron-tipped staff that could be called a spear.[80]

---

78 Stifter's reference to "Horneck" repeats the false attribution by sixteenth-century scholar Wolfgang Lazius. In Franz Grillparzer's influential drama *König Ottokars Glück und Ende* (1825), the character "Ottokar von Horneck" (not the king), delivers a single, yet all-the-more famous speech in the third act "in praise of Austria" (*Lob auf Österreich*).

79 Cf. Zumbusch, "Der Gang der Geschichte," 243.

80 HKG 5.1, 18–19 [16].

Stifter presents a series of historical objects one might find in a cultural-historical museum: the split earthen jug; the mug plained with flowers; the crossbow and staff or spear. In addition to the external focalization, such descriptions reveal an uncertainty about names and functions: is a staff with an iron tip really a spear? The negative affirmation, "He was wearing nothing else," raises a further question essential to historical reconstruction: How does one determine whether the relics of the past represent all the objects that once existed? The poor innkeeper's simple cloth tunic almost certainly did not survive the seven centuries to find its way into a nineteenth-century historical museum, but perhaps his gray jacket, "made of coaser material," did. Or perhaps only the more robust leather doublets and helmets survived along with the ceramics, wood, and iron objects that constitute the most common materials of any medieval historical collection.

The novel's hero, Witiko, also wears a leather helmet and doublet. The epithet "Ledermann" ("leather man") underscores the close assocation of the character and the material of his armor.[81] The first description of the protagonist, moreover, focuses on the material qualities of the leather armor he wears:

> Das Haupthaar konnte nicht angegeben werden; denn es war ganz und gar von einer ledernen Kappe bedeckt, welche, wie ein Becken von sehr festem und dickem Stoffe gebildet, so daß ein ziemlich starker Schwerthieb kaum durchzudringen vermochte, [... sie] lag ihm vielmehr dicht an, und wurde unter dem Wamse geborgen, welches von gleichem Leder den ganzen Oberkörper knapp umhüllte. [...] Von dem nämlichen Leder schien auch die Beinbekleidung des Reiters. All dieses Leder war ursprünglich mattgelb gewesen, und wiewohl man nicht verkennen konnte, daß große Sorgfalt auf seine Erhaltung und Reinigung angewendet worden sei, so mußte man doch zugeben, daß es nicht mehr neu sei, und Spuren von Wetterschäden und ausgetilgten Flecken zeigte.
>
> You couldn't tell the color of his hair since it was completely covered by a bowl shaped leather helmet made of such a firm solid material that even a rather strong sword's blow couldn't penetrate it. [It] was tucked into a narrow doublet of the same type of leather which covered the entire upper part of his body [...] The armor on his legs seemed to be of the same type of leather. This armor had originally been light yellow but even though it was obvious that the greatest care had been taken as regards to its maintenance and cleanliness, one had to admit that it was no longer new and showed signs of weather damage as well as gouges (*ausgetilgten Flecken*).[82]

Witiko's leather armor fits him snugly like a second skin. In the context of Stifter's goal of recreating "historical life" and the organic metaphors used to describe this effort – which I discuss in detail below – the material "leather"

81 HKG 5.1, 68. Frye somewhat oversells the epithet by translating it as "Sir Knight in Leather," even if the moniker is as memorable as the descriptions of the armor itself [49].
82 HKG 5.1, 16 [14–15].

has special significance. As the material of armor, it has the same status as any other object that is part of a historical milieu. But as animal skin, leather also refers both metonymically and metaphorically to the living bodies who once wore it.[83] Leather is a kind of fossilized skin, a relic of past life. And as we learn here, it is susceptible to weathering and wear, even when one takes "the greatest care" to keep it clean and maintain it. What is more, its "life" becomes uniquely legible in the spots that resist removal and must be "gouge[d]" out ("ausgetilgten Flecken"). Its suppleness thus renders it a medium for recording its own history.

Still, the cleanliness of Witiko's leather armor contrasts sharply with the "very dirty" leather doublets of the men he will soon meet at the inn. With this contrast, Stifter establishes a correlation to the "conservative" values Witiko represents and embodies in the care of his armor – i. e. the values of an art conservator! – and the moral rectitude he represents as the novel's hero. In the second volume of the novel, those men at the inn make use of the aforementioned "crossbow" in an attempt to murder and rob Witiko and his men.[84] The episode thus distinguishes the one and true "leather man" Witiko from his leather doubles, who, not coincidently, appear as a pair from the start. In general, the dress worn by the novel's characters is recognizable, or even explicitly described, as a *combination* of a limited set of elements.[85] Thus the young riders whom Witiko soon encounters on his way to Prague:

> Sie waren alle jung und in schöne Farben gekleidet. Die meisten waren röthlich oder rothbraun, die anderen grün. Sie hatten gleichfalls Federn auf den Hauben, Reigenfedern, Hahnenfedern und dergleichen. Manche hatten ein Hüfthorn, alle hatten ein Schwert, und einige trugen eine oder gar mehrere Jagdlanzen.
>
> They were all young and dressed in beautiful colors. Most were in reddish or russet, the others in green. They too had feathers in the helmets, heron, cock and the like. Some had hunting horns, all had swords, and a few were carrying one or several hunting spears.[86]

At times, moreover, Stifter even forgoes the operation of recombination and appears to lead his readers directly into the historical museum itself, or at least, the "storerooms" of a historical castle. When Witiko enters a (historical) castle, he

---

83 As parchment, leather represents the older and more stable medium of writing than the paper of modern books. See the discussion of Witiko as a "piece of paper" below.

84 HKG 5.2, 163 f. [296 f.].

85 Cf. Heiko Christians and Oliver Kohns, "Poetik der Einfachheit. Stifters 'Witiko,'" *Wirkendes Wort* 55, no. 3 (2005): 389.

86 HKG 5.1, 64 [46].

does not merely conduct whatever business he has there, he also takes time to look around. Such visits provide an opportunity for Stifter to describe both the historical castle architecture, with its chapel, dining room, reception halls of varying sizes, private living and bedrooms, military and servant quarters. After the architectural tour comes the tour of the armory:

> Sie gingen in die Rüstkammern, in denen die Waffen der Vertheidigung und des Angriffes waren, Panzerhemden, Schilde, Helme, Lederrüstungen, Schwerter, Lanzen, Bogen, Pfeile, Köcher, Armbruste, und ähnliche Dinge. Sie gingen in den Raum der Wurfgeräte, Schutzkörbe, Flechtwerke und anderer Mittel. Dann besahen sie die Pferde und die anderen Thiere in den Ställen, und die Räume der Vorräte.
>
> They went to the armory that contained a variety of weapons and armor: coats of mail, shields, helmets, leather armor, swords, lances, bows, arrows, quivers, cross bows and such. They also went to where catapults, missile containers, material for revetements and other devices were stored. Then they inspected the horses and the other animals in the stables as well as the storerooms.[87]

"Clubs axes swords," indeed! If the occasional naming of historical objects can contribute to the historical atmosphere of a novel, Stifter's systematic catalogs create a fog so thick as to alienate readers and obscure their historical meaning. The scraps of tunic and leather armor have become an entire tableau of historical detail. Stocked high in such a storeroom of historical knowledge, these objects neither evoke the specific milieu of Balzac's inn nor do they produce any "reality effect."

Nonetheless, such descriptions suggest a parallel between the historical experience provided by a cultural-historical museum and the reconstruction of history in the historical novel. The concept of milieu underlies and motivates both forms of reconstruction. Indeed, Stifter's novel describes natural milieus in the same manner as the interiors and storerooms of the castle:

> Witiko aber blieb in seinem Gemache, ging an ein Fenster, und sah auf das Land Österreich hinab. Der Wald des Berges trug nicht wie die Wälder, die an den Ufern der jungen Moldau stehen, die dunkle Tanne, die grüne Buche, die leuchtende Birke, den langarmigen Ahorn, und die Eibe, die Ulme, die Esche, die Erle und den zackigen Wachholder, sondern gleichartiges Laubwerk der Buche, Esche, Eibe, des Haselstrauches und das Dämmer der Föhre und andern Nadelgehölzes.
>
> Witiko remained in the chamber, stepped to a window and gazed down at the Land of Austria. The trees on the mount weren't like those on the banks of the Upper Moldau: dark pines, green beeches, gleaming birches, long limbed maples, elms, ashes, alders, and

87 HKG 5.1, 214 [139].

> pointed junipers; these were trees whose foliage were similar: beeches, ashes, yews, hazel bushes and the darnkess of spruces and other evergreens.[88]

The internal focalization on the protagonist's look out the window leads here to a unique form of ekphrasis. The extended list of tree species communicates not only what Witiko sees and thus, how the forest looks, but also his observation of a difference. The forest of this lowland "mount" near Vienna is not the same as the forest of his birthplace near the source of the Moldau river. The difference resides in the particular combination of species – some like beech and ash grow in both places. A forest is thus not just a forest, it is a milieu, or what we now call an ecosystem. The "geography of plants," to employ Alexander von Humboldt's epochal botanical term, indicates the geography of Austria, and describes Witiko's voyage from his home in the Bohemian forest to the castle of the Margravess Enzelin in Lower Austria. In the same way, the geography of cultural-historical objects not only indicates the specific milieus of the novel's settings, from roadside inn to the palaces of Prague and Vienna, from monastery to castle, but also, more generally, the historical world of the twelfth century. It thus makes sense that Stifter felt he had to visit Prague to study its topography before writing a key battle scene set there.[89] By studying the ruins of Wittinghausen or the landscapes surrounding Prague as they are now, the author can imagine, describe, and thus "reconstruct" such milieus as they once were.

Stifter's "scientific" approach to writing a historical novel not only carried the risk of descriptive monotony, however, but also the possibility of anachronism. The first readers and critics of *Witiko* appealed to cultural history to judge the novel's historical accuracy. An anonymous critic from the *Augsburger Allgemeinen Zeitung* (July 1, 1865) even employs the term "Culturgeschichte," while others, such as Hieronymous Lorm, discuss the "Culturmomente" in the novel.[90] With a better knowledge of cultural history, Stifter might have avoided the *faux pas* of placing a fork in the hands of his twelfth-century hero. The fact that this little mistake bothered Stifter so greatly is evidence of how important archeological knowledge was to his poetic process.[91] When he received word about the critical response to the first volume in 1865, the detail about Witiko's fork clearly nagged at him: "Es sind im 1[ten] Bande Unrichtigkeiten, die mich jezt

88 HKG 5.2, 235 [341].

89 Stifter travelled to the Bohemian capital in June 1865, cf. commentary in HKG 5.4, 193–194.

90 Reprinted in: HKG 5.4, 275–276.

91 One can also connect the seemingly insignificant fork to Stifter's poetic preference for "small things," cf. Christians and Kohns, "Poetik der Einfachheit," 392ff.

quälen" [there are inaccuracies in the first volume that torture me now].[92] While Stifter tried to treat it as a bagatelle, "Ich mußte herzlich lachen" [I just had to laugh about it],[93] the anachronism of the fork even causes him to question the very possibility of an accurate historical account: "aber wer kann jede Faser alten Lebens so geben, wie sie gewesen ist, wir leben in einer neuen Zeit, und gerade im Kleinsten gleitet ohne Bewußtsein des Schreibers oft eine Neuheit ins Werk, die zuweilen kindisch ist" [but who can render every fiber of ancient life, as it was, we live in a new age, and especially when it comes to the smallest things, a childish anachronism sometimes slips into the work without the author's awareness].[94] And self-doubt gives way to a broad critique of historians more generally: "Ob die Alterthümler ganz recht haben, wenn sie feststellen, wie einmal alles gewesen ist, weiß ich nicht" [I don't know whether the antiquarians are quite right when they determine how everything once was].[95]

The phrase "every fiber of ancient life" belongs to a series of metaphors that Stifter employs in his letters that conceptualize the poetics of the historical novel as the reconstruction of organic remains, the restoration of living tissue. These metaphors express the revival of both history and epic poetry as a form of geohistorical reconstruction. They recall the classic hermeneutic topos of "dead letter" vs. "living spirit," but give it anatomical contours. Because it is garnered from the dead letter of books, historical "knowledge" does not suffice when the goal is to reproduce historical "life:" "Aber zu dem ist nicht das historische Wissen allein genug, dies gäbe nur ein hölzernes Gerippe, sondern das historische Mitleben, dieses gibt den Gestalten Fleisch und Blut" [Moreover, historical knowledge on its own is not enough, this would only present a wooden skeleton, while it is historical life that gives the figures flesh and blood].[96] From a geohistorical perspective, the "skeleton" or *Gerippe* is hardly even a metaphor. Archeological research is always a study of skeletons and other ruins. The paradoxical combination of the skeleton metaphor with the predicate "wooden," moreover, suggests that cultural artifacts and natural products can be studied in the same way. The metaphor of anatomical reconstruction thus serves as a global analogy for historical "work" as such: "Wenn Sie wüßten, welche Arbeit es ist, aus den alten Urkunden und Dokumenten den Körper des Mittelalters zu con-

92 PRA 21, 2.
93 PRA 21, 3.
94 PRA 21, 2.
95 PRA 21, 3.
96 PRA 18, 169.

struiren" [If you only knew how much work it is to construct the body of the middle ages out of old records and documents].[97]

When Georges Cuvier reconstructed the skeleton of the wooly mammoth, "imaginations" followed him to lend this body flesh, blood, and fur. No one expected a skeleton would come to life, except perhaps in science fiction novels from *Frankenstein* to *Jurassic Park*. Indeed, Stifter's aim "to construct the body of the Middle Ages" resembles such science-fiction fantasies, but, despite moments of frustration, the author maintains that his historical revival is realistic. In his antiquarian enthusiasm, Stifter tends to forget the incompatibility of his own age with the deep historical past. This incompatibility may appear in the form of anachronisms, but it also characterizes the text on the level of form. With *Witiko*, Stifter felt that his poetic process had undergone a radical transformation: "Darum ist mir öfter, als hätte ich früher nur geschwärmt und dichtete jezt. Die Griechen hatten das Wort ποιεω (ich mache) für Dichten. Wie bezeichnend! Gestalten muß man machen, nicht Worte" [It thus often feels as if I had only been daydreaming and now I am writing poetry. The Greeks had the word ποιεω (I make) for writing poetry. How true! One must make figures, not words].[98] This statement of poetic intent encapsulates the dilemma Stifter faces as soon as he endeavors to transform the "words" of historical documents into "figures," into a "body" or "life." By translating ποιεω as "I make," Stifter reinvigorates the metaphorical meaning of "poetry" as a plastic art, but appears to forget that the "material" of poetry is words.[99]

To make a plastic "figure" (*Gestalt*) in poetry, one must appeal to the signifying power of language. Language can only signify, however, through a material signifier and immaterial signified, a letter and a spirit. Yet if metaphors of the body predominate in Stifter's descriptions of his project for a novel, these metaphors become more "spiritual" in his defense of the finished work. Confronted with the irritating example of the anachronistic fork, for example, Stifter writes:

> Was wäre mit dem herrlichen Niebelungenliede, das in Atilas Zeit spielt und lediglich Sitten, Kämpfe etc des zwölften Jahrhunderts bringt, also ein ganzer Verstoß gegen die Zeit um 700 Jahre ist, wie meines Witikos Messer und Gabel auch um einige hundert Jahre verstoßen mögen. Wäre nur jeder Kummer, jede Sorge, jede Pein von meinem Gemüthe ferne, daß so viel Luft des Alterthums und der Dichtung in meinem Werke wehte, als in meinem Wesen liegt, mehr vermag ja kein Sterblicher zu geben, als er hat.

97 PRA 17, 303.

98 PRA 19, 266.

99 Stifter's at times willfully naive theorization of literary semiotics has been explored at length by Geulen, *Worthörig wider Willen*, and Begemann, *Die Welt der Zeichen*.

> What about the majestic song of the Nibelungs, which takes place in the time of Attila and yet presents customs, battles etc. from the twelfth century, thus is a total breach against the period by about 700 years, just as my Witiko's knife and fork are also a breach of several centuries. If only every worry, care and scruple were far from my heart that as much air of antiquity and poetry would waft through my work as in in my being; a mortal cannot give more than he has.[100]

What allows the modern author to write about the past is the "air of antiquity and poetry." The site of this historical atmosphere is not around or between all the objects, people, and places named in the work, but in the author's very "being" (*Wesen*). The metaphor of air thus exposes conceptual weaknesses of the body metaphor. How can poetry provide "flesh and blood" or the historical "living-with" (*Mitleben*) if not via signs? And what good is a sign without its signified, the immaterial spirit of meaning?

The Epic and the Novel: Simulating Orality in Prose

By examining metaphors of body and spirit in Stifter's commentary of his work on *Witiko*, one can thus deconstruct the novel's claims to historical reconstruction. Yet even in its failure to reproduce historical life, *Witiko* does breathe with the life of its own historical period. In defense of the work, the critic Hieronymous Lorm thus praises above all its "peculiar spirit:"

> Allein bei dem eigenthümlichen Geist, der das Werk trägt und durchdringt, unter dem Licht einer Verklärung der Dinge, welches mehr Wahrheit, ja selbst mehr geschichtliche Wahrheit sehen läßt, als ein Spiegel der gemeinen Wirklichkeit, wird der kunstfreundliche Leser [...] die Versündigungen an der ordinären Beschaffenheit der Nebendinge dem Dichter [...] bereitwillig verzeihen.
>
> But in consideration of the peculiar spirit that holds up the work and fills it, in light of the transfiguration of things which shows more truth, indeed, more historical truth than the mirror of common reality, the artistically-minded reader will readily forgive the poet his sins against the vulgar composition of secondary things.[101]

Lorm appeals here to the central category of German poetic realism, *Verklärung*. Paradoxically, the aestheticization and idealization of poetic realism produces "more truth" and even "more historical truth" than "common reality." The distinction between some form of higher truth or beauty and a common, ordinary

100 PRA 21, 3.
101 For Lorm, a fork is clearly a *Nebending*. Cited in HKG 5.4, 275.

reality is a frequent topos of realist poetics in the mid-nineteenth century. Stifter's own famous formulation of the poetics of *Verklärung* is "real reality" or "wirkliche Wirklichkeit."[102] If Lorm praises Stifter's accomplishment for its adherence to contemporary novel poetics, his assessment also makes quite clear what is at stake in such idealizations: the things of the past (*Dinge*), so important to the geohistorical reconstruction Stifter envisions, become trivialities or "secondary things" (*Nebendinge*).

If we imagine Nietzsche reading Lorm's review of *Witiko*, and apply his views on truth, historical truth, and life to the claims of poetic realism, one thing is clear: the philosopher would hardly buy the argument that there is "more historical truth" in a novel because it is full of poetic spirit. Yet without being fooled, the "artistically-minded reader" Nietzsche would likely view the obfuscation of historical truth in the novel as one of its "advantages" to present life. In Nietzschean terms, the movement from "trivial things" such as a fork to the "peculiar spirit" of the novel implies a shift from antiquarian to monumental history. In both cases, however, Nietzsche clearly places a priority on present life over historical truth.

Stifter sees matters differently. Even though the author defends his "being" (*Wesen*), and his present, "mortal" life, he claims that this being is full with the air of not only "poetry," but also "antiquity." According to Nietzsche's concept of "monumental" history, one advantage of history is that it provides models. For Stifter, the model for the combination of poetry and antiquitity is the "majestic song of the Nibelungs." In consideration of Stifter's enthusiastic embrace of the ancient Greeks and their *poiesis*, moreover, one can extend this model to the genre of the epic. Indeed, Stifter explicitly calls the historical novel "das Epos in ungebundener Rede" [the epic in prose],[103] though he does not turn to the epic so as to imitate it precisely, but instead pursues a poetological strategy that elevates his own poetics of the historical novel through a favorable comparison to the epic: "Die Weltgeschichte als ein Ganzes, auch die ungeschriebene eingerechnet, ist das künstlerischeste Epos" [World history as a whole, also including unwritten history, is the most artistic epic].[104] If history is itself an epic, then the highest aims of literature can be achieved by writing history, that is, by

102 The phrase itself appears in the novella *The Descendants*, HKG 3.2, 65. The passage associates "real reality" with both landscape painting and epic poetry.

103 PRA 19, 283.

104 PRA 19, 224. I do not read this programmatic statement to mean that history already has a literary or narrative form, namely "the epic," unlike Cornelia Zumbusch, "Der Gang der Geschichte," 230. Instead, like others ca. 1800, Stifter imagines "the epic" as the literary form that is so simple that it does not falsify history.

writing the historical novel. If the inverse also holds, the "most artistic epic" guarantees privileged access to history, indeed to unwritten history.

Stifter's positive appropriation of the epic genre left traces on the work itself. First, the subtitle designates the genre as "story" (*Erzählung*), not "novel" (*Roman*). The war plot and the characterization of the protagonist as a hero are appropriate for the epic genre. The hero himself exemplifies the virtue of "constancy" or *Stetigkeit*, an aesthetic quality also attributed to the epic style and hexameter form. Questions of form and style, however, also present difficulties to a reading of *Witiko* as an epic. While catalogs and, more generally, a paratactic style are typical for the epic, Stifter's novel is not written in verse. One could argue that verse form is not a necessary feature of a (systematic) definition of the epic genre. But the question of verse or prose inevitably raises the question of *historical* differences in the poetics of the genre, and to the key aspect of the nineteenth-century understanding of the epic. Beginning in the late eighteenth century, discussions of the epic centered on the distinction between the epic and the novel. From Friedrich Schlegel to Wilhelm von Humboldt to G.W.F. Hegel and Georg Lukàcs, there was widespread consensus among German-language critics and theorists that the novel succeeded the epic. Within the broader generic trinity (*Gattungstrias*), both novel and epic (*Epos* or *Epopöe*) are forms of *Epik*, as opposed to *Lyrik* and *Drama*. In the context of his epochalization of his philosophy of history, Hegel thus codifies the succession as an irreversible replacement, calling the novel "die moderne bürgerliche Epopöe" [modern bourgeois epic].[105]

As Heiko Christians has shown, however, the "epic" did not simply disappear circa 1800, nor was it simply replaced by the novel.[106] Instead, a tradition of literary texts, from Goethe's *Hermann und Dorothea* to Peter Handke's *Die Wiederholung*, and including Stifter's *Witiko*, as well as literary theories of the novel, notably Georg Lukàcs's, invoke the epic tradition in more or less emphatic ways, from allusion to imitation and modernization. In the case of *Witiko*, Stifter chooses the modern prose form, but he cites the verse origins of the epic, for example, in the chapter titles. The first of these titles is "Es klang wie der Gesang von Lerchen" [It sounded somewhat like the song of larks].[107] While the lark's song (*Ge-*

---

105 G.W.F. Hegel, *Vorlesungen über die Ästhetik III*, ed. Eva Moldenhauer and Karl Markus Michel, Werke vol. 15 (Frankfurt a.M.: Suhrkamp, 1986), 392. Cf. Blasberg's account of Stifter's borrowings from contemporary theories of the epic, *Erschriebene Tradition*, 304–311.

106 Heiko Christians, *Der Traum vom Epos. Romantik und politische Poetik in Deutschland (1750-2000)* (Freiburg im Breisgau: Rombach, 2004), 29–52.

107 HKG 5.1, 13 [13]. Frye adds the "somewhat" to produce an assonance in imitation of the German internal rhyme.

sang) alludes metaphorically to the "song" of the epic poet, the rhyme of *klang/ Gesang* produces the "sound" (*Klang*) of song or oral poetry. The "like" of the simile, however, insists on the difference: while this novel is like epic song, it is still prose. The chapter title thus upholds the distinction between novel and epic, even as it maintains the fusion of the two sides as a rhetorical possibility, or as Christians writes, as a "dream of the epic." In this reading, the combination "epic novel" becomes possible not by modifying the definition of epic, which would involve eliminating characteristic features such as verse form, but through the inclusion of the epic within the flexible form of the modern novel. In this way, the epic becomes another and perhaps the most important feature of medieval history that Stifter aims to reconstruct or revive in *Witiko*.

Cornelia Blasberg makes a strong case that Stifter identifies the epic primarily with the Nibelung tradition. Not only can one find countless intertextual parallels between the two texts, but the novel also describes the creation of the *Nibelungenlied*, which it suggests was written by Witiko's school friends, the Ritter vom Kürenberge and Heinrich von Ofterdingen.[108] This fictional attribution recalls a thesis by Anton von Spaun.[109] Spaun's title, *Heinrich von Ofterdingen und das Nibelungenlied. Ein Versuch, den Dichter und das Epos für Österreich zu vindicieren* (*Henry of Ofterdingen and the Song of the Nibelungs. An Attempt to Claim the Poet and the Epic for Austria*, 1840) reveals the nationalist motivations of literary history circa 1850, but also points to the complexity with which Stifter responds to his mentor's strong claim. By including Kürnberger in addition to Ofterdingen in the novel, Stifter not only cites a competing authorship theory by the Viennese literary scholar Franz Pfeiffer, but also echoes the theory of collective authorship of oral epics first popularized by Friedrich August Wolf.[110] By fictionalizing the disputed authorship of the *Song of the Nibelungs*, Stifter's novel thus writes itself into literary history, as both a literary-historical and literary work.

As Stifter's letter quoted above demonstrates, the antiquity of epic poetry resides not only in its (verse) form, but also in the content of its plot: the saga of the Nibelungs was written in the twelfth century, but recounts events from the "time of Attila," or the fifth century. Not only does this fanciful dating bring the mythic time of Sigfried and Krimhild into historical time, it also places the nineteenth-century author into a lineage of epic poets. Indeed, the parallelism

108 HKG 5.3, 340. Cf. Blasberg, *Erschriebene Tradition*, 249–327.

109 Stifter wrote an obituary for Spaun in which he praises the scholar's writings and Nibelungen thesis in particular, HKG 8.1, 47–51. See also Moriz Enzinger, *Adalbert Stifter und die altdeutsche Literatur* (Vienna: Österreichische Akademie der Wissenschaften, 1961).

110 Blasberg, *Erschriebene Tradition*, 298–299.

of the dates implies that history and literary history repeat themselves in a cycle of seven hundred years.

The title of Christians's study, *The Dream of the Epic*, acknowledges that the ancient epic tradition was or at least was perceived as irretrievably lost; Blasberg's title, *Self-Written Tradition*, points to the central role of writing in this cultural-historical process. From Herder's theorizing on *Volkspoesie* ("people's poetry") to Friedrich August Wolf's highly influential thesis that the Homeric poems were the product of oral traditions, the discourse on the epic ca. 1800 had concluded that the distinction epic/novel largely corresponded with the difference between oral and written poetry. In this context, the question of verse or prose is not merely one of literary form, but speaks to the conditions of production and hence the very conditions of possiblity of epic poetry. It is no coincidence that Stifter "resolved" the question of the authorship of the *Nibelungs* by relating it to the idea of the collective production of poetic song.

As I have tried to show in this study, however, geohistoricism is an attitude towards the past that pays particular attention to its physical traces. In the reinvigorated metaphor of the "book of nature" or the earth's "room of records," moreover, these physical traces appear as writing. While geohistory is limited to the interpretable traces in the geological record, within these limits, the purview of history also expanded beyond the bounds of what was previously imagined possible. As Nietzsche's critique of historicism shows, however, the resistence to the pressure of historical overload was immediate, and perhaps necessary. If writing is the gateway to history, in other words, the living still need oral poetry.

At the very least, the desire for or specter of oral poetry produces the most significant rift in the poetic conception of Stifter's historical novel. Although the author of *Witiko* may enthusiastically embrace the idea of "making" historical "figures" instead of "words," he seems to forget that "words" are the stuff, the material of literature. A passage from an earlier letter is revealing in this regard:

> Ich habe das Gerippe von *sechs* Bänden zusammen getragen, nehmlich zwei Romane. Alle Scenerie und Handlung ist geordnet, der Organismus ist gegliedert, und es fehlt nur die Textirung, die hier das unendlich leichtere ist, als bei der Fantasieerzählung, weil hier das Wesentlichste und Schwierigste die Herausfindung des Körpers des Mittelalters ist, der aber, wenn er einmal gefunden ist, sofort für sich selber dasteht, und keines Kleides mehr bedarf, als ers schon anhat.

> I have assembled the bones of six volumes, namely two novels. All the scenes and plot are ordered, the organism is articulated, and all that is missing is the filling with text, which is incomparably easier than in an invented story because here the most fundmental and most difficult thing is to discover the body of the Middle Ages, which, once it has been found,

immediately stands for itself, and does not need any other dress than the one it already has on.[111]

While the metaphor of "articulation" (*Gliederung*) has become standard in German as the term for "outline," here its origins in the concept of the organism and its skeleton are still clear. What also becomes clear in this letter, however, is the fact that "filling with text" (*Textirung*) is the process required to give or restore life to this organic skeleton. The metaphorical pair body/dress (*Körper/Kleid*) implies that the act of textual restoration does not add anything that was not always already there.[112] Stifter's historical *inventio* is thus a form of finding (*Finden*) and not imaginary invention (*Erfinden*): "ich muss sie [die Form des Romans] finden, nicht *er*finden" [I must find the form of the novel, not invent it]. Not surprisingly, Stifter sees this "finding" as analogous to the discoveries of the natural scientist: "Das Finden macht mir aber oft große Freude, wie dem Naturforscher, wenn er unbekannte aber längst vorhandene Erscheinungen entdekt" [But finding often gives me great joy, like it does to the natural scientist when he discovers phenomena that have long existed but have remained unkown].[113]

It is worth recalling at this point that Viollet-le-Duc, in his article "On Restoration," includes philology among the historical sciences that followed Georges Cuvier on his geological path: "Next comes the philologist, who discovers the origin of European languages, all issuing from the same source."[114] In Viollet-le-Duc's analogy, the history of words resembles the history of things: words are understood as a succession of historical forms. Historical grammar focuses its attention on the material side of language, its sound changes, and its "inner structure," the morphology that belongs to the order of the signifier.[115] Just as it is possible to retrace the history of language, so too with the history of literature.

111 PRA 17, 302–303.

112 As Zumbusch notes, the metaphor of "dress" (*Kleid*) suggests here a metaphor of text as textile, a reading that resonates with Stifter's frequent use of the dress metaphor to describe the relationship between beauty and truth, "Der Gang der Geschichte," 236 f.

113 PRA 19, 266.

114 Viollet-le-Duc, "On Restoration," 14. See Chapter 3.

115 This inner structure is the "anatomy" of language: "Jener entscheidende Punkt aber, der hier alles aufhellen wird, ist die innre Structur der Sprachen oder die vergleichende Grammatik, welche uns ganz neue Aufschlüsse über die Genealogie der Sprachen auf ähnliche Weise geben wird, wie die vergleichende Anatomie über die höhere Naturgeschichte Licht verbreitetet hat" [the decisive point that will elucidate everything here is the inner structure of languages or comparative grammer, which will provide us with new insights into the genealogy of languages in a similar way to how comparative anatomy has shed light on higher natural history]; F. Schlegel, "Über die Sprache und Weisheit der Inder. 1808," 137.

The most famous model of literary restoration that develops in the nineteenth century is textual criticism. Through his famous theory of the *Urtext*, Karl Lachmann develops a method that closely corresponds with Viollet-le-Duc's definition of restoration.[116] In Lachmann's editions of medieval manuscripts, each manuscript is treated as a fragment, as partial evidence of an underlying *Urtext* – a presumed original text that "may never have existed."[117]

The reconstruction of languages and literatures does not "revive" them, however. The reception of medieval literature ca. 1800 involves a wide variety of strategies for making old texts available to modern readers. For Romantic writers such as Achim von Arnim and Clemens Brentano, who both translate and compose their own "Old German Songs" (*Des Knaben Wunderhorn. Alte Deutsche Lieder,* 1805), or Ludwig Tieck, who offers translations or "new renderings" (*neu bearbeitet*) of Swabian Minnesong (*Minnelieder aus dem schwäbischen Zeitalter,* 1803), the presentation of a simulated oral folk tradition is an explicit poetic strategy. Tieck justifies this strategy through an idealistic appeal to "One Poetry," which expresses itself from the "earliest times" into the "furthest future" as an "indivisible Whole."[118]

Tieck's contemporary, Friedrich Heinrich von der Hagen, published a translation of the *Song of the Nibelungs* in 1807. As a law student, Hagen heard lectures by Friedrich August Wolf in Halle and later August Wilhelm Schlegel in Berlin, and occupies a rather idiosyncratic position between professional philology and Romantic poetry. He praises Tieck's Minnesong translations, for example, but suggests that the poet's renderings are too "arbitrary." In the afterword to this work, Hagen thus stakes out a middle-ground between translation and philological edition, which he calls "Erneuung" [modernization].[119] His arguments

---

116 Cf. Timpanaro, *The Genesis of Lachmann's Method*, and Harald Weigel, *"Nur was du nie gesehn wird ewig dauern:" Carl Lachmann und die Entstehung der wissenschaftlichen Edition* (Freiburg: Rombach, 1989).

117 Blasberg, *Erschriebene Tradition*, 256, contrasts Lachmann's approach with that of Jacob Grimm, who insists on the uniqueness of every historical artifact, a position that John Ruskin will later occupy in criticizing Viollet-le-Duc.

118 "Denn es giebt doch nur Eine Poesie, die sich selbst von den frühesten Zeiten bis in die fernste Zukunft, mit den Werken die wir besitzen, und mit den verlohrnen, die unsre Phantasie ergänzen möchte, so wie mit dem künftigen, welche sie ahnden will, nur ein unzertrennliches Ganze ausmacht." Ludwig Tieck, *Minnelieder aus dem schwäbischen Zeitalter, neu bearbeitet und herausgegeben* (Berlin: Realschulbuchhandlung, 1803), 2.

119 Friedrich von der Hagen, *Der Nibelungen Lied* (Berlin: Unger, 1807), 477. Cited in Blasberg, *Erschriebene Tradition*, 257. Later, having become a professor in Breslau, Hagen produced a philological edition "in the original language:" *Der Nibelungen Lied in der Ursprache mit den Lesart-*

in favor of this method are particuarly relevant for Stifter's *Witiko*. Like his contemporary philologists, Hagen imagines the historical linguistic change as "stätige Fortbildung" [constant development] and "lebendige Umwandlung" [living transformation]. Such terms recall Viollet-le-Duc's concept of "uninterrupted filiation" or indeed, Stifter's analogy between the "succession" of natural historical forms and artistic styles. Like Viollet-le-Duc, moreover, Hagen makes a distinction between this smooth, organic development and the actual effects of history. In Hagen's view, an editor must restore a text "zu ihrer vollendeten Bildung" [to a state of completion], but this does not mean, "daß man diese Werke so aufstellen sollte, die ganze Zeit hindurch, unter allen wirklichen Umständen, im Umlauf geblieben wären, zu uns herübergekommen sein möchten" [that one should present it as if had reached us after having remained the whole time in circulation, under actual circumstances].[120] For such "actual circumstances" are a constant source of corruption, of copying and printing errors, which Hagen judges to be as inauthentic as the acts of destruction and renovation of buildings that Viollet-le-Duc will later seek to remove. Hagen's concept of the continuity of linguistic development thus effaces historical difference. Because modern German has developed organically out of earlier forms, his "modernization" of the medieval text is not a translation, but itself a form of organic literary development or "Fortbildung," while the centuries that divide him from the original *Nibelungs* can simply be "jumped over."[121]

Stifter at no point conceives his historical novel as a direct restoration of any one medieval text, even the *Nibelungs*. As Blasberg argues, the *Song of the Nibelungs*, whose themes of family strife and revenge play a major role in *Witiko*'s composition, is merely the central knot in a web of intertextuality that links Stifter's historical sources to Romantic novels such as Novalis's *Henry of Ofterdingen*.[122] Blasberg also shows how Stifter's complex inscription of the historical transformations of his medieval *Ur-Text* in *Witiko* speaks both to the novel's "unheard-of modernity" and help explains its significant "conceptual breaks."[123]

To conclude, I would like to discuss one of these "conceptual breaks," which I would argue runs through the project as a whole, from its intial conception to the final acts of writing. This break runs along the faultline between medieval

---

en aller Handschriften und Erläuterungen der Sprache, Sage und Geschichte (Breslau: Josef Mar, 1820).

120 Ibid.

121 "so daß der Zwischenraum bis zu ihrer vollendeten Bildung, der sie nur hätte verunstalten können, mit einemmal gleichsam übersprungen wird." Ibid.

122 Blasberg, *Erschriebene Tradition*, 316.

123 Ibid. 327; 270.

speech and modern writing. Lachmann's textual-critical reconstruction and Hagen's "modernization" represent two strategies for resolving this conflict. Variants of both strategies are at work in the writing of *Witiko:* on the one hand, Stifter remains faithful to the "words" of the medieval period by simulating orality and reproducing the philological evidence of his historical studies; on the other hand, the overarching poetic strategy remains loyal to descriptive and narrative modes of the realist novel. The language of *Witiko* is no more similar to the Middle High German of the twelfth century than it is in the verse form of the *Song of the Nibelungs*, but despite such "modernization," Stifter's text simulates orality in a manner that, to say the least, serves the interests of historical mimesis more than narrative diegesis.

Here is an example from Wladislaw's council in Prague, where Wladislaw questions witnesses to his diplomatic exchange with the rival Konrad:

> "Ich habe die Worte gehört," sprach Zwest.
> "Ich habe die Worte gehört," sprach Wecel.
> "Ich habe die Worte gehört," sprach Zdeslaw.
> "Ich habe die Worte gehört," sprach Bohuslaw.
> "Ich habe die Worte gehört," sprach Casta.
>
> "I heard his words," said Zwest.
> "I heard his words," said Wecel.
> "I heard his words," said Zdeslaw.
> "I heard his words," said Bohuslaw.
> "I heard his words," said Casta.[124]

This is the minimal form in which the historical names, found in an ancient chronicle, can take shape, become *Gestalten*, and "come to life." The repetition of the phrase "I heard his words" underscores the performative function of these words, which have no propositional content. Throughout *Witiko*, one can find examples of highly schematized, indeed ritualized speech.[125] In this extreme case, speech is reduced to a roll-call, to a performance of bearing witness to the speech of another. This repetition is the necessary consequence of the poetology that considers history "das künstlerischeste Epos" [the most artistic epic]. Stifter continues: "und wenn Theile davon als Dichtung genommen werden, so sind sie am schönsten, wenn sie einfältiglich heraus gehoben und aus dem Munde des mitlebenden Volkes erzählt werden. Der Gelehrte und der heutige Dichter verder-

124 HKG 5.2, 316–317 [389–390], translation modified.

125 Cf. Kai Kauffmann, "Mündliche Rede," 127–131; and Alice Bolterauer, "Das Ritual in der Moderne. Überlegungen zu Adalbert Stifters 'Witiko,'" *Dogilmunhak* 68, no. 4 (1998): 65–86.

ben nur daran" [and when parts of it are taken as poetry, then they are the most beautiful when they are excerpted in a simple way and are told from the mouth of the people who lived it. The scholar and the modern poet can only fail at such a task].[126] Indeed, many readers of Stifter's novel attribute this "failure" (*verderben*) to the monotonous, repetitive style on display in this passage. In particular, the repetition from the direct, reported speech, which has a justifiably ritual character, also appears in the novel's narrative discourse. In translating the above citiation, Wendell Frye finds such contamination so unbearable that he sacrificies accuracy for rhetorical variation: Stifter's fivefold repeated "sprach" becomes "attested [...] added [...] said [...] spoke [...] spoke."

The spoken word, from the mouth of living people, is the medium where the material and meaning of history meet, but only when it is recorded. After the witnesses attest to Konrad's fatal declaration of defiance, Wladislaw orders: "Schreibe die Worte in das Pergament" [Write these words on the parchment].[127] The order to write thus establishes the order of speech as a political order. As the double meaning of the term "protocol" implies, the order of ritualized speech expresses a hierarchy among speakers.[128] As Kai Kauffmann argues, this act of writing is one of several instances – including the registry of battle casualties and the survey and registration of landholdings – that solidify a new regime of "centralized admistrative power."[129] Wladislaw's justified and notarized decision to repress Konrad's rebellion in battle belongs to one of several real and fictive texts whose recording is described in the novel.[130] Stifter's text thus bears witness to its own origins in textual traditions, even as it insists on the orality of the medieval life. Out of texts, it reconstructs the lived creation of the texts that served as the basis of the reconstruction in the first place.

In this way, the novel itself adds complexity to Stifter's description of his poetological process, where first the "organism" is structured as a "skeleton" and then "filled with text" (*Textirung*). *Witiko* expresses this same complexity on a metaphorical level, as well. If its messenger-hero acts as a kind of notary, who attests, by listening, to the accuracy and authenticity of messages, he can do so because he is as neutral as "a piece of paper:"

126 PRA 19, 224.

127 HKG 5.2, 317 [390].

128 Zumbusch, "Der Gang der Geschichte," 246.

129 For the other two instances, see HKG 5.2, 100, and 5,3 83; cited in Kauffmann, "Mündliche Rede," 134–135.

130 Cf. Blasberg, *Erschriebene Tradition*, 281; and Moriz Enzinger, "Der Plan zu 'Wok' und die Rosenberger-Triologie," in *Gesammelte Aufsätze zu Adalbert Stifter* (Vienna: Österreichische Verlagsanstalt, 1967), 253.

> [...] daß mich die Versammlung die Beschlüsse anhören lasse, damit ich dem Herzoge nichts Unreines und Geschändetes bringe. Ich selber komme nicht in Betracht, so wenig, wie ein Stücklein Papier, darauf eine hohe Hand eine Zeile geschrieben hat, die man findet, und achtet.
>
> That the assembly permit me to hear the decision so I don't bring the duke anything false, harmful, or slanderous. I myself am not important, nor do I merit any more consideration than a piece of paper on which a noble lord has written a line which you find and respect.[131]

As a hero, Witiko embodies the values of faithfulness and loyality, the value of being able to give one's word and hold fast to it. He embodies the values, in other words, of a culture of orality.[132] Yet the only truly faithful and accurate messenger is writing. Witiko sees writing as the necessary condition for the "fame" of his contemporaries, already the stuff of "songs and stories," to last into a future long after their deaths: "Und die Menschen haben die Kunst erfunden, ihre Worte in Buchstaben zu legen, die dauern, und durch diese Erfindung und durch das, was noch erfunden werden wird, lebt der Ruhm fort, wenn die, welche Großes verübt haben, längst schon vor dem Throne Gottes sind" [And men have learned the art of transposing their words into letters that last, and by this invention and by that which is still to be invented, fame will live long after those who did the great deeds are before the Throne of God].[133] As the *telos* of the "great deeds" in *Witiko* is thus to become writing, it is fitting that its style most often resembles a "transcript" of oral communication.[134] And while the "invention" still to come may well be the movable type that gave birth to the modern world, and the modern historical novel, it may also refer to the technique of Stifter's *Witiko:* to recover the words of the past by imagining that they have always been text after all.

131 HKG 5.1, 119 [81].
132 Kauffmann, "Mündliche Tradition," 134.
133 HKG 5.3, 274 [557]. Translation modified.
134 Kauffmann, "Mündliche Tradition," 138–140.

Epilogue

In April 2019, the spire of the Notre-Dame Cathedral in Paris collapsed in flames. Commentators sought consolation: the spire was not original, but a later addition, so what was lost was not authentic thirteenth-century architecture, they reasoned, but a nineteenth-century fake. With that, Viollet-le-Duc thus returned to public debates about cultural heritage and memory, about medieval and modern architecture and the relative merits of restoration, conservation, and other forms of reconstruction. There is little consolation, and a great deal of irony, in the argument that the fire might "only" have destroyed Viollet-le-Duc's restorations. The desire to see the original, thirteenth-century structure emerge from the flames, purified of their later, barbaric supplements, revives the spirit of the very architect accused of such inauthentic additions.

Another irony, also widely discussed, points to a more fundamental paradox of the conservation of monuments. Notre-Dame was undergoing restoration work when the fire broke out. Whether or not the construction work was the cause of the fire, the coincidence recalls the commonplace criticism of restoration: by changing the material of the past, restoration inevitably destroys the past it seeks to conserve. At the same time, the elemental violence of the fire reminds us that historical monuments are always subject to the destructive forces of nature. The conservation of monuments thus requires constant intervention, even if the architects of such actions reject the label "restoration."

The fire may have started by accident, but in the long history of the cathedral, such accidents inevitably occur. The geological reconstruction of the past reveals earth history as one of constant and ineluctable destruction. Geohistoricism thus intertwines positive and negative moments of destruction and reconstruction. In the face of such destruction, the geohistorical restorations reconstruct the past out of its material remnants.

Of course, my study of the geological underpinnings of Viollet-le-Duc's historicism does not legitimize his restoration choices, nor does it provide guidance for the current restoration of the Notre-Dame Cathedral, a project that the president of France has established as an urgent five-year plan. As the example of Notre-Dame demonstrates, each restoration project has its own specific cultural and political context. The restorers of Notre-Dame may well choose an anti-historicist path, imitating Norman Foster's post-modern rebuilding of the Berlin Reichstag rather than a complete restoration, like that of the *Frauenkirche* in Dresden. As these, as well as other examples – such as the façade restorations of the Berlin *Stadtschloss* [City Palace] or the Altstadt in Frankfurt – show, the desire to restore the monuments of the past is as strong in our present as it

https://doi.org/10.1515/9783110685534-007

was nearly two hundred years ago. If anything, the paradigm of historical conservation established in the second quarter of the nineteenth century has gained in appeal. Once associated with a romantic fascination with Gothic art and the project of modern nation building, today's historical conservationism is global and universal. When the Islamic State destroys ancient pagan monuments in Palmyra or the Taliban destroy Buddhist statues in Afghanistan, the immediate reaction of the international community – supported institutionally by organizations like UNESCO – is to restore what has been lost.

While the concept of geohistoricism does not explain the specific motivations of individual restoration projects, it does offer an alternative narrative for understanding the relationship between conservation and modernity. In the typical narrative, modernity accelerates social change to such an extent that it provokes historicism as a reaction. Here, one can understand historicism both as a sharpened sense of historicity, including the historicity of the present, and as aesthetic historicism, the fascination with history or a particular historical epoch. In this view, historicists and historical conservationists wish to preserve the past against a modern present, which in its tireless march into the future, proceeds by forgetting. Nietzsche criticizes this form of historicism or antiquarianism to the extent that it impedes forgetting necessary for life – even as he shows sympathy for the historicist's inherent and "untimely" skepticism of the present day. Stifter likewise appeals to historical conservationism as a protective measure against modern "indifferentism."

The alternative narrative of geohistoricism suggests that the consciousness of historical temporality has more far-reaching consequences than a social critique of industrial modernity. While eighteenth-century debates touted the historical specificity of the project of Enlightenment, the historicization of all things that occurred across disciplines ca. 1800, including the emergent field of geology, effected an entirely different historical attitude in both breadth and quality. When Viollet-le-Duc credits Cuvier with introducing this attitude, he is merely acknowledging the radicality of this change. Before historical geology, humans wrote of dramas that they might have collectively called history; they did not recognize that the ground on which they stood, the natural stage of human life, was itself subject to historical change. As Stifter's reflections in the catacombs demonstrate, this geohistorical change can itself have erosive effects. The erosion of all material things to rubble, sand, and dust is its own form of destructive indifference.

At the same time, the knowledge of earth history is a product of reconstructions that provide new methods for the reconstructions and restorations of human history and its artefacts as well. It is perhaps not necessary to follow these paths as Viollet-le-Duc or Stifter does. Today's archeologists carefully reconstruct past worlds, but acknowledge the various ways in which such reconstructions remain in the virtual world of scientific knowledge, digital visualization, or experimental analogy. Architects and artisans are building a medieval castle in Guédelon, for example, that in many respects resembles Viollet-le-Duc's reconstruction of the castle at Pierrefonds, but is conceived purely as a model of medieval architecture and building practices, not as a "restoration" in any sense of the term. Yet, as in the nineteenth century, every refinement in archeological reconstruction returns to the original collaboration of geology and history that characterizes geohistoricism. Aerial photography and three-dimensional scans may improve the accuracy of the imaginary reconstruction of past forms through drawing, and chemical analysis may provide more precise dating than speculations about the organic evolution of architectural forms, but the fundamental alliance between geology and other historical sciences remains intact.

Nineteenth-century authors like Viollet-le-Duc and Stifter are acutely aware of the connections between geological and other forms of historical reconstruction. While Viollet-le-Duc considers geology the initiator of a new historical method across a variety of disciplines, Stifter sees earth history as a central node in the network of sciences, and hence as an appropriate stand-in for science in general. At the turn of the twentieth century, as the fascination with geology itself arguably dissipates, geohistorical thought continues to take on new forms. For example, Marcel Proust, who consulted Viollet-le-Duc's *Dictionary of French Architecture* frequently as a source of historical knowledge about the Middle Ages, compares his literary method of reconstructing personal memory to the architect's mode of "restoration." Similarly, Sigmund Freud imagines excavating the psyche as if it were an archeological dig, only to recognize, like Proust and indeed Stifter and Viollet-le-Duc before him, that the act of excavation and remembrance produces something new, which may never before have existed.

In Hubert Damisch's work on Viollet-le-Duc, moreover, it becomes clear that structuralism is also indebted to the organicism and anatomical functionalism of the nineteenth century. It was a work of abstraction to understand, with Damisch, that the imbrications of architectural structures suggest a general system of relation and difference, or, with Georges Canguilhem – to cite a more prominent example – that the milieu is a concept of pure relation. Readers of Claude Lévi-Strauss's *Tristes tropiques* will note the author's claim to a fascination with geology, which leads him to relate the structuralist concept of synchrony to the geological stratum.

Foucault's poststructuralist metaphor of "the archeology of knowledge" is only the first of more recent such transfers. While Foucault distinguishes in his essay "Of Other Spaces" between the nineteenth century's fascination with time, and the twentieth century's turn to space, the name he chooses for his own historical method reminds us of the great achievement of nineteenth-century geology and archeology of making time visible through space.

The archeology of media, finally, echoes nineteenth-century metaphors of the letters or records room of earth history. When the media are stone, they are always the messengers of their own history. Media archeology emerges from the recognition that this history matters as much as whatever other letters may be engraved on stone, or printed on the transformed remnants of an ancient forest.

It is neither possible nor desirable to trace the lineage of all these varied forms of geohistoricism to Viollet-le-Duc, Stifter, Georges Cuvier, or Charles Lyell. What the latter names represent is rather a decisive moment in the historicization of nature, when earth history not only freed itself from ancient cosmogonies, but also from the theoretical speculations of a Newtonian Age. Earth history became history, a complex one-way street of contingent events, and at the same time, it exploded the temporal limits of that history. Stifter and Viollet-le-Duc are not only sensitive witnesses to this new historical episteme, but also active participants in the construction of a new restorative poetics. Each sees in the ruins of (earth) history the remnants of a complete world, and each longs to see that world reconstructed. Viollet-le-Duc responds with an elaborate theory of architectural restoration that combines the creativity of his imagination with the rationality of scientific correlation. In his literary imagination, Stifter reflects, through narrative structure and the metaphorical language of literature, on the specific historical conditions of his own desire to reconstruct the past.

Such reflections reveal the paradoxes and aporias of restoration practices. Both Viollet-le-Duc and Stifter demonstrate an awareness of the fact that restorations are always late interventions in a vain battle against the unstoppable forces of erosion. In the mid-nineteenth century, the vast depths of earth history served as a reminder of the precariousness of all human endeavors. Today, the concept of the Anthropocene reverses Stifter's notion of human history as a mere "interpolation" in earth history, or rather, it turns this interpolation into a short episode of earth history with a grotesquely large and permanent impact. The effects of human industry over a mere two centuries thus become comparable to the asteroid collision that caused the last great extinction event and the demise of the dinosaurs, sixty-six *million* years ago. The growing consciousness of human impact on the global environment in the Anthropocene has shifted the meaning, but not diminished the significance of geohistoricism. Viollet-le-Duc's organicism results from an analogy that models the products of art and culture according to the rules that govern works of nature. The Anthropocene also elides the boundaries between culture and nature in its recognition that human culture now governs over the totality of natural spaces on earth. In this sense, there is no more nature and no more “environment.” As a result, we engage today in the “restoration” of natural ecosystems more readily than in the restoration of works of art, which we prefer to “conserve,” keeping visible the material traces of time.

In addition to his theoretical, historical “restoration” of Mont Blanc, Viollet-le-Duc imagined the “aménagement des terroirs” [land management] of the surrounding areas, to protect the natural beauty of the area. Stifter also presents land management as the ultimate horizon of his restorative fantasies. The agricultural utopia in *Indian Summer* resembles recent experiments in organic farming and agricultural tourism more than it does any real agrarian past. The criterion of authenticity, however, is less relevant to natural restorations than to artistic ones. Instead of authenticity, the success of natural conservation efforts depends on *Nachhaltigkeit* or sustainability, a concept from eighteenth-century forestry. From this perspective, it becomes apparent that, from *Castle Crazy* to *Indian Summer* and *The Descendants*, Stifter's fictions not only ask whether what has been lost can be restored (*wiederhergestellt*), but also what form of poiesis or construction (*Herstellung*) is restorable and hence sustainable. In the novella *Castle Crazy*, the Green Fichtau, the living household, holds this potential, while in *The Descendants*, Stifter pessimistically assesses the devastating consequences of a culture built on the clearing of forests and the draining of wetlands.

In *Indian Summer*, finally, Stifter imagines an ideal ecology, but concedes its limits within the frame of earth history. As the center of the idyllic, post-estival world created by the Freiherr von Risach, the Rose House represents a perfect *oikos*, a site of restoration, but also of cyclical renewal. Such renewal echoes the novel's cyclical narrative structure, organized around the yearly visits to the House to admire its magnificent wall of roses. Heinrich marries Natalie and the couple becomes the future heirs of Risach's estate, but the older man sets conditions for the inheritance, or more precisely, his condition is that the transfer remain unconditional. In his wisdom, he forbids that the heirs treat the Rose House as a museum. Excessive respect for inherited goods annuls the testamentary act of transfer. As grateful heirs to the ideal world Risach has recreated, the heirs resist, and insist that, at the very least, they will preserve the alder brook that provides a sustainable source of beautiful wood for the furniture studio. Here Risach concedes, “daß ihm nichts angethan werde von Geschlecht zu Geschlecht, bis seine Reste vermodern, oder ein Wolkenguß ihn von seiner Stelle feget” [that nothing may be done to change it from generation to generation until its remnants rot or a cloudburst sweeps it away].[1] From generation to generation, we can pass on the values of conservation, until one day, we recognize that decay and erosion are as essential to the earth's own regeneration as sedimentation and growth.

1 HKG 4.3, 280 [477].

Bibliography

Primary Sources

Balzac, Honoré de. *La Comédie humaine.* Edited by Pierre-Georges Castex. 12 vol. Paris: Gallimard, 1976 – 1981.

Balzac, Honoré de. *The Wild Ass's Skin.* Edited by Patrick Coleman. Translated by Helen Constantine. Oxford: Oxford UP, 2012.

Buffon, Comte de (Leclerc, Georges-Louis). "Du granit." In *Histoire naturelle des minéraux*, 1:96 – 129. Paris: Imprimerie Royale, 1783.

Burckhardt, Jacob. *Über das Studium der Geschichte.* Edited by Peter Ganz. Munich: Beck, 1982.

Burckhardt, Jacob. "Über geschichtliches Studium." In *Gesammelte Werke*, Vol. IV. Basel: Schwabe, 1956.

Cuvier, Georges. *Recherches sur les ossements fossiles des quadrupèdes.* Paris: Deterville, 1812.

Diderot, Denis. *Oeuvres complètes*, vol. 16, ed. Else Marie Bukdahl, Michel Delon, and Annette Lorenceau. Paris: Hermann, 1990.

France, Anatole. "Pierre Nozière." In *Œuvres*, edited by Marie-Claire Bancquart, 3:487 – 645. Paris: Gallimard, 1991.

Goethe, Johann Wolfgang von. "Granit II." In *Aufsätze, Fragemente, Studien zur Naturwissenchaft im Allgemeinen; Die Schriften zur Naturwissenschaft (Leopoldina)*, edited by Dorothea Kuhn and Wilhelm Engelhardt, 11:10 – 14. Weimar: Böhlau, 1970.

Goethe, Johann Wolfgang von. "On Granite (1784)." In *The Essential Goethe*, edited by Mathew Bell, 913 – 915. Princeton: Princeton UP, 2016.

Goethe, Johann Wolfgang von. *Sämtliche Werke.* Edited by Friedmar Apel et al. 40 vol. Vol. 10. Frankfurt a. M.: Deutscher Klassiker Verlag, 1987 – 2013.

Goethe, Johann Wolfgang von. *The Collected Works.* Edited by Jane K. Brown. Translated by Jan van Heurck and Krishna Winston. Vol. 10. Princeton: Princeton UP, 1995.

Hagen, Friedrich von der. *Der Nibelungen Lied.* Berlin: Unger, 1807.

Hall, Sir James. *Essay on the Origin, History, and Principles of Gothic Architecture.* London: W. Bulmer, 1813.

Hegel, G.W.F. *Vorlesungen über die Ästhetik III. Werke*, vol 15. Edited by Eva Moldenhauer and Karl Markus Michel. Frankfurt a. M.: Suhrkamp, 1986.

Kant, Immanuel, *Kritik der praktischen Vernunft. Kritik der Urtheilskraft. Akademie Ausgabe V* (Berlin: Königlich Preußische Akademie der Wissenschaften, 1913).

Lyell, Charles. *Principles of Geology, Volume 1.* Chicago: U Chicago P, 1990.

Mercier, Louis-Sébastien. *Tableau de Paris. Nouvelle édition* [1782]. Reprint. Vol. 1. Geneva: Slatkine, 1979.

Mohs, Friedrich. *Die Charaktere der Klassen, Ordnungen, Geschlechter und Arten; oder, Die Charakteristik des naturhistorischen Mineralsystemes.* Dresden: Arnoldische Buchhandlung, 1821.

Nietzsche, Friedrich. *Digital Critical Edition of the Complete Works and Letters*, based on the critical text by G. Colli and M. Montinari, Berlin/New York: De Gruyter, 1967–, edited by Paolo D'Iorio. http://www.nietzschesource.org/#eKGWB.

https://doi.org/10.1515/9783110685534-008

Nietzsche, Friedrich. "On the Utility and Liability of History for Life." In *Unfashionable Obervations, The Collected Works of Friedrich Nietzsche 2*, translated by Richard T. Gray, 83–167. Stanford: Stanford UP, 1995.

Nietzsche, Friedrich. *The Portable Nietzsche*, trans. Walter Kaufmann. New York: Viking Penguin, 1977.

Nietzsche, Friedrich. "Unzeitgemäße Betrachtungen II. Vom Nutzen und Nachtheil der Historie für das Leben." In *Die Geburt der Tragödie; Unzeitgemäße Betrachtungen I–IV; Nachgelassene Schriften 1870–1873*, edited by Giorgio Colli and Mazzino Montinari, 243–334. Kritische Studienausgabe 1. Munich, Berlin, New York: DTV/De Gruyter, 1988.

Novalis. *Henry of Ofterdingen. A Romance.* Edited by John Owen. Cambridge, MA: Cambridge Press, 1842.

Novalis. *Werke*, ed. Gerhard Schulz. Munich: Beck, 1969.

Poe, Edgar Allan. "The Cask of Amontillado." In *The Collected Works of Edgar Allan Poe. Vol. III: Tales and Sketches*, edited by Thomas Ollive Mabbott, 1252–1266. Cambridge, MA: Harvard, 1978.

Ruskin, John. *The Works of John Ruskin.* Edited by E.T. Cook and A. Wedderburn. Vol. 26. 39 vols. London: George Allen, 1903.

Sammons, Jeffrey L., ed. *German Novellas of Realism: Stifter, Droste-Hülshoff, Gotthelf, Grillparzer, Mörike.* Vol. 1. New York: Continuum, 1989.

Schlegel, Friedrich. "Über die Sprache und Weisheit der Inder. 1808." In *Kritische Friedrich-Schlegel-Ausgabe*, edited by Ernst Behler, 8:105–433. Paderborn: Schöningh, 1975.

Stifter, Adalbert. *Castle Crazy and Maroshely.* London: R. Bentley, 1851.

Stifter, Adalbert. *Indian Summer.* Translated by Wendell Frye. 4th ed. Bern: Peter Lang, 2009.

Stifter, Adalbert. *Sämmtliche Werke.* Edited by August Sauer. 23 vols. Prague: J.G. Calve, 1901–1939 [PRA].

Stifter, Adalbert. *Werke und Briefe: Historisch-Kritische Gesamtausgabe.* Edited by Alfred Doppler, Wolfgang Frühwald, and Hermann Kunisch. 10 vols. Stuttgart: Kohlhammer, 1978 ff. [HKG].

Stifter, Adalbert. *Witiko.* Translated by Wendell Frye. 2nd ed. Bern: Peter Lang, 2006.

Tieck, Ludwig. *Minnelieder aus dem schwäbischen Zeitalter, neu bearbeitet und herausgegeben.* Berlin: Realschulbuchhandlung, 1803.

Viollet-le-Duc, Eugène-Emmanuel. *Dictionnaire raisonné de l'architecture française du XIe au XVIe siècle.* 10 vols. Paris: B. Bance and A. Morel et Cie., 1854–1868.

Viollet-le-Duc, Eugène-Emmanuel. *Dictionnaire raisonné du mobilier français de l'époque carlovingienne à la Renaissance.* 6 vols. Paris: B. Bance and A. Morel et Cie., 1858–1875.

Viollet-le-Duc, Eugène-Emmanuel. *Histoire d'un dessinateur: comment on apprend à dessiner.* Paris: J. Hetzel, 1879.

Viollet-le-Duc, Eugène-Emmanuel. *Learning to Draw: Or the Story of a Young Designer.* Translated by Virginia Champlin. New York: G.P. Putnam's Sons, 1881.

Viollet-le-Duc, Eugène-Emmanuel. *Le Massif du Mont Blanc, étude sur sa constitution géodésique et géologique sur ses transformations et sur l'état ancien et moderne de ses glaciers.* Paris: J. Baudry, 1876.

Viollet-le-Duc, Eugène-Emmanuel. *Lettres d'Italie (1836–1837) adressées à sa famille.* Edited by Geneviève Viollet-le-Duc. Paris: Léonce Laget, 1971.

Viollet-le-Duc, Eugène-Emmanuel. *On Restoration*. Translated by Charles Wethered. London: Sampson Low, Marston Low, and Searle, 1875.
Viollet-le-Duc, Geneviève. *Les Viollet-le-Duc. Histoire d'une famille. Documents et correspondance*. Geneva: Slatkine, 2000.

Secondary Literature

Amiel, Christiane. "Viollet-le-Duc et la montagne." In *Eugène Viollet-le-Duc: regards croisés. Actes du colloque de Carcassonne 5 et 6 juin 2014*, edited by Jean-Pierre Piniès. Carcassonne: Garae Hésiode, 2014.
Anderson, R. Lanier. "Friedrich Nietzsche." In *The Stanford Encyclopedia of Philosophy*, edited by Edward N. Zalta. Stanford: Metaphysics Research Lab, Stanford U, 2017. https://plato.stanford.edu/archives/sum2017/entries/nietzsche/.
Arburg, Hans-Georg von. "Neues von der Narrenburg? Stifters Architekturen zwischen Historismus und Neuem Bauen." In *Figuren der Übertragung*, edited by Michael Gamper and Karl Wagner, 109–133. Zürich: Chronos, 2009.
Arnold-De Simine, Silke. "Musealisierungsphänomene im Werk Adalbert Stifters." In *Ordnung-Raum-Ritual: Adalbert Stifters artifizieller Realismus*, edited by Sabina Becker and Katharina Grätz, 41–68. Heidelberg: Winter, 2007.
Attanucci, Timothy. "Atmosphärische Stimmungen: Landschaft und Meteorologie bei Carus, Goethe und Stifter." *Zeitschrift für Germanistik* 24, no. 2 (2014): 282–295.
Attanucci, Timothy. "'The Gentle Law' of Large Numbers: Stifter's Urban Meteorology." *Monatshefte* 112, no. 1 (2020): 1–19.
Auerbach, Erich. *Mimesis: Dargestellte Wirklichkeit in der abendländischen Literatur*. 10th ed. Tübingen: Francke, 2001.
Auzas, Pierre-Marie. *Eugène Viollet le Duc, 1814–1879*. Paris: Caisse Nationale des Monuments Historiques, 1965.
Auzas, Pierre-Marie. "Préface." In *Actes du colloque international Viollet-le-Duc Paris 1980*, edited by Pierre-Marie Auzas, 7–9. Paris: Nouvelles Editions Latines, 1982.
Bachelard, Gaston. *The Formation of the Scientific Mind: A Contribution to a Psychoanalysis of Objective Knowledge*. Manchester: Clinamen, 2002.
Bajohr, Hannes. "Keine Quallen. Anthropozän und Negative Anthropologie." *Merkur* 73, no. 804 (2019): 63–74.
Bajohr, Hannes, ed. *Der Anthropos im Anthropozän: Die Wiederkehr des Menschen im Moment seiner vermeintlich endgültigen Verabschiedung*. Berlin/Boston: De Gruyter, 2020.
Bakhtin, Mikhail. "Forms of Time and of the Chronotope in the Novel." In *The Dialogic Imagination. Four Essays*, edited by Michael Holquist, translated by Caryl Emerson and Michael Holquist, 84–258. Austin: U Texas P, 1981.
Banitz, Edgar. "Das Geologenbild Adalbert Stifters." In *Gestaltung, Umgestaltung. Festschrift zum 75. Geburtstag von Hermann August Korff*, edited by Joachim Müller, 206–238. Leipzig: Koehler & Amelang, 1957.
Baridon, Laurent. "Anatomie comparée et pensée évolutionniste dans la théorie et la pratique architecturale de Viollet-le-Duc." In *L'architecture, les sciences et la culture de*

l'histoire au XIXe siècle, edited by François Loyer, 73–81. Saint-Etienne: P.U. de Saint-Etienne, 2001.

Baridon, Laurent. *L'imaginaire scientifique de Viollet-Le-Duc*. Paris: Harmattan, 1996.

Bark, Irene. *"Steine in Potenzen": Konstruktive Rezeption der Mineralogie bei Novalis*. Tübingen: Niemeyer, 1999.

Barthes, Roland. "The Reality Effect." In *The Rustle of Language*, 141–148. Berkeley: U California P, 1989.

Baßler, Moritz, Christoph Brecht, Dirk Niefanger, and Gotthart Wunberg. *Historismus und literarische Moderne*. Tübingen: Max Niemeyer, 1996.

Beer, Gillian. *Darwin's Plots: Evolutionary Narrative in Darwin, George Eliot and Nineteenth-Century Fiction*. Revised edition. Cambridge: Cambridge UP, 2000.

Begemann, Christian. *Die Welt der Zeichen: Stifter-Lektüren*. Stuttgart: Metzler, 1995.

Begemann, Christian. "Metaphysik und Empirie: Konkurrierende Naturkonzepte im Werk Adalbert Stifters." In *Wissen in Literatur im 19. Jahrhundert*, edited by Lutz Danneberg and Friedrich Vollhardt, 92–127. Tübingen: Niemeyer, 2002.

Begemann, Christian, and Davide Guiriato, eds. *Stifter-Handbuch. Leben – Werk – Wirkung*. Stuttgart: Metzler, 2017.

Belgum, Kirsten L. "High Historicism and Narrative Restoration." *The Germanic Review* 67, no. 1 (1992): 15–25.

Benjamin, Walter. "On Some Motifs in Baudelaire." In *The Writer of Modern Life: Essays on Charles Baudelaire*, edited by Michael Jennings, 170–212. Cambridge, MA: Harvard UP, 2006.

Benjamin, Walter. "The Paris of the Second Empire in Baudelaire." In *The Writer of Modern Life: Essays on Charles Baudelaire*, edited by Michael Jennings, 46–143. Cambridge, MA: Harvard UP, 2006.

Bercé, Françoise. *Viollet-le-Duc*. Paris: Éditions du patrimoine, 2013.

Blasberg, Cornelia. *Erschriebene Tradition: Adalbert Stifter oder das Erzählen im Zeichen verlorener Geschichten*. Freiburg i.B.: Rombach, 1998.

Blumenberg, Hans. *Die Lesbarkeit der Welt*. 2nd ed. Frankfurt a. M.: Suhrkamp, 1983.

Blumenberg, Hans. *Paradigmen zu einer Metaphorologie*. 2nd ed. Frankfurt a. M.: Suhrkamp, 1999.

Böhme, Hartmut. "Das Steinerne. Anmerkungen zur Theorie des Erhabenen aus dem Blick des 'Menschenfremdesten.'" In *Das Erhabene: Zwischen Grenzerfahrung und Größenwahn*, edited by Pries, Christine, 119–142. Weinheim: VCH, Acta Humaniora, 1989.

Bolterauer, Alice. "Das Ritual in der Moderne. Überlegungen zu Adalbert Stifters 'Witiko.'" *Dogilmunhak* 68, no. 4 (1998): 65–86.

Borchmeyer, Dieter. "Stifters Nachsommer – eine restaurative Utopie?" *Poetica (Munich)* 12 (1980): 59–82.

Boudon, Françoise. "Le réel et l'imaginaire chez Viollet-le-Duc: les figures du 'Dictionaire de l'architecture.'" *Revue de l'Art* 58–59 (1983): 95–114.

Braungart, Georg. "Apokalypse in der Urzeit. Die Entdeckung der Tiefenzeit in der Geologie um 1800 und ihre literarischen Nachbeben." In *Zeit-Zeitenwechsel-Endzeit. Zeit im Wandel der Zeiten, Kulturen, Techniken und Disziplinen*, by G. Leinsle and Jochen Mecke, 107–120. Regensburg: Universitätsverlag, 2000.

Braungart, Georg. "Der Hauslehrer, Landschaftsmaler und Schriftsteller Adalbert Stifter besucht den Gletscherforscher Friedrich Simony. Hallstatt, im Sommer 1845." In *Bespiegelungskunst. Begegnungen auf den Seitenwegen der Literaturgeschichte*, edited by Georg Braungart, Friedrmann Harzer, Hans Peter Neureuter, and Gertrud Rösch, 101–118. Tübingen: Attempto, 2004.

Bressani, Martin. *Architecture and the Historical Imagination: Eugène-Emmanuel Viollet-Le-Duc, 1814–1879*. Farnham UK: Ashgate, 2014.

Bressani, Martin. "Opposition et équilibre: Le rationalisme organique de Viollet-le-Duc." *Revue de l'Art* 112 (Spring 1996): 28–37.

Bressani, Martin. "Science, histoire et archéologie. Sources et généalogie de la pensée organiciste de Viollet-le-Duc." Université de Paris-Sorbonne, 1997.

Brevern, Jan von. *Blicke von Nirgendwo: Geologie in Bildern bei Ruskin, Viollet-le-Duc und Civiale*. Munich: Fink, 2012.

Briese, Olaf. "'Typen' oder 'Individuen'? Personennamen in literarischen Sittenstudien um 1840." *Interférences littéraires/Literaire interferenties* 8 (May 2012): 135–151.

Buckland, Adelene. *Novel Science: Fiction and the Invention of Nineteenth-Century Geology*. Chicago: U Chicago P, 2013.

Bühler, Benjamin. *Ecocriticism: Grundlagen – Theorien – Interpretationen*. Stuttgart: Metzler, 2016.

Byrd, Vance. "Beautiful Form? Vienna and the Viennese and Stifter's Urban Sketches." *Interférences Littéraires/Literaire Interferenties* 8 (May 2012): 155–169.

Canguilhem, Georges. "Le vivant et son milieu." In *La connaissance de la vie*, 165–198. Paris: J. Vrin, 1965.

Catalano, Gabriella. "Griechische Spuren in Stifters *Nachsommer*." In *Graecomania: Der europäische Philhellenismus*, edited by Gilbert Heß, Elena Agazzi, and Elisabeth Décultot, 99–115. Berlin: De Gruyter, 2009.

Christians, Heiko. *Der Traum vom Epos. Romantik und politische Poetik in Deutschland (1750–2000)*. Freiburg i.B.: Rombach, 2004.

Christians, Heiko, and Oliver Kohns. "Poetik der Einfachheit. Stifters 'Witiko.'" *Wirkendes Wort* 55, no. 3 (2005): 389–403.

Crutzen, Paul J., and Eugene F. Stoermer. "The Anthropocene." *IGBP Newsletter* 41 (2000): 17–18.

Crutzen, Paul J. "Geology of Mankind." *Nature* 415, no. 3 (2002): 23.

Damisch, Hubert. "Introduction." In *Viollet-le-Duc, L'architecture raisonnée, extraits du "Dictionnaire raisonné de l'architecture,"* by Eugène-Emmanuel Viollet-le-Duc, 9–26. edited by Hubert Damisch. Paris: Hermann, 1964.

Damisch, Hubert. "Les 'Entretiens sur l'architecture' ou du structuralisme au fonctionnalisme." In *Actes du colloque international Viollet-le-Duc Paris 1980*, edited by Pierre-Marie Auzas, 93–99. Paris: Nouvelles Editions Latines, 1982.

Dehio, Georg, and Alois Riegl. *Konservieren, nicht restaurieren. Streitschriften zur Denkmalpflege um 1900*. Edited by Marion Wohlleben and Georg Mörsch. Reprint. Berlin/Boston: Birkhäuser, 1988.

Deleuze, Gilles, and Félix Guattari. *A Thousand Plateaus: Capitalism and Schizophrenia*. Translated by Brian Massumi. 2nd ed. Minneapolis: U Minnesota P, 1987.

Domandl, Sepp. "Die philosophische Tradition von Adalbert Stifters 'Sanftem Gesetz.'" *Vierteljahresschrift des Adalbert Stifter-Instituts des Landes Oberösterreich* 21, no. 3/4 (1972): 79–103.
Downing, Eric. "With Wandering Steps and Slow: Schwellenkunst in Adalbert Stifter's *Granit*." *DVjs* 94 (2020): 69–86.
Dünne, Jörg. *Die katastrophische Feerie: Geschichte, Geologie und Spektakel in der modernen französischen Literatur*. Konstanz: Konstanz UP, 2017.
Dürbeck, Gabriele, and Urte Stobbe, eds. *Ecocriticism: Eine Einführung*. Cologne: Böhlau, 2015.
Dvorák, Max. "Denkmalpflege in Österreich [1911]." *Österreichische Zeitschrift für Kunst und Denkmalpflege* 28, no. 3 (1974): 131f.
Eck, Caroline van. *Organicism in Nineteenth-Century Architecture: An Inquiry into Its Theoretical and Philosophical Background*. Amsterdam: Architectura & Natura Press, 1994.
Eckert, Jörn. *Der Kampf um die Familienfideikommisse in Deutschland: Studien zum Absterben eines Rechtsinstitutes*. Frankfurt a. M.: Peter Lang, 1992.
Ehlbeck, Birgit. *Denken wie der Wald: Zur poetologischen Funktionalisierung des Empirismus in den Romanen und Erzählungen Adalbert Stifters und Wilhelm Raabes*. Bodenheim: Philo, 1998.
Endres, Johannes, ed. *Friedrich Schlegel-Handbuch: Leben – Werk – Wirkung*. Stuttgart: Metzler, 2017.
Enzinger, Moriz. *Adalbert Stifter und die altdeutsche Literatur*. Vienna: Österreichische Akademie der Wissenschaften, 1961.
Enzinger, Moriz. *Adalbert Stifters Studienjahre (1818–1830)*. Innsbruck: Österreichische Verlagsanstalt, 1950.
Enzinger, Moriz. "Der Plan zu 'Wok' und die Rosenberger-Triologie." In *Gesammelte Aufsätze zu Adalbert Stifter*, 235–254. Vienna: Österreichische Verlagsanstalt, 1967.
Finance, Laurence de, and Jean-Michel Leniaud, eds. *Viollet-le-Duc: les visions d'un architecte*. Paris: Norma, 2014.
Fleming, Paul. *Exemplarity and Mediocrity: The Art of the Average from Bourgeois Tragedy to Realism*. Stanford, CA: Stanford UP, 2008.
Fliethmann, Axel. *Stellenlektüre: Stifter, Foucault*. Tübingen: Max Niemeyer, 2001.
Flöring, Karl. *Die historischen Elemente in Adalbert Stifters "Witiko."* Gießen: Münchow, 1922.
Foucault, Michel. *The Order of Things: An Archaeology of the Human Sciences*. London: Routledge, 2005.
Freud, Sigmund. "Eine Schwierigkeit der Psychoanalyse." In *Gesammelte Werke*, 7th ed., 12:3–12. Frankfurt a. M.: Fischer, 2005.
Frey, Pierre A., and Lisa Grenier, eds. *Viollet-le-Duc et la montagne*. Grenoble: Grénat, 1993.
Frodl, Walter. *Idee und Verwirklichung. Das Werden der staatlichen Denkmalpflege in Österreich*. Vienna: Böhlau, 1988.
Gamper, Michael. "Adalbert Stifter: 'Der Nachsommer' (1857)." In *Literatur und Wissen: ein interdisziplinäres Handbuch*, edited by Roland Borgards, Harald Neumeyer, Nicolas Pethes, and Yvonne Wübben, 370–374. Stuttgart: Metzler, 2013.
Gamper, Michael. *Elektropoetologie: Fiktionen der Elektrizität 1740–1870*. Göttingen: Wallstein, 2009.

Gamper, Michael. *Masse lesen, Masse schreiben: eine Diskurs- und Imaginationsgeschichte der Menschenmenge 1765–1930.* Munich: Fink, 2007.

Gamper, Michael. "Meteorologie/Wetter." In *Stifter-Handbuch. Leben – Werk – Wirkung*, edited by Christian Begemann and Davide Guiriato, 253–257. Stuttgart: Metzler, 2017.

Gamper, Michael. *Vorsicht. Emergenz eines Dispositivs der Moderne.* Berlin: Alpheus, 2015.

Gamper, Michael, and Karl Wagner, eds. *Figuren der Übertragung: Adalbert Stifter und das Wissen seiner Zeit.* Zürich: Chronos, 2009.

Garrard, Greg. *The Oxford Handbook of Ecocriticism.* Oxford, New York: Oxford UP, 2014.

Gerlach, Franziska Frei. "Die Macht der Körnlein. Stifters Sandformationen zwischen Materialität und Signifikation." In *Die Dinge und die Zeichen: Dimensionen des Realistischen in der Erzählliteratur des 19. Jahrhunderts*, edited by Sabine Schneider and Barbara Hunfeld, 109–122. Würzburg: Königshausen & Neumann, 2008.

Geulen, Eva. "1848, October 11: Tales of a Collector." In *A New History of German Literature*, edited by David E. Wellbery and Judith Ryan, 587–592. Cambridge, MA: Harvard UP, 2004.

Geulen, Eva. *Worthörig wider Willen: Darstellungsproblematik und Sprachreflexion in der Prosa Adalbert Stifters.* Munich: Iudicium, 1992.

Gilbert, Felix. *History: Politics or Culture. Reflections on Ranke and Burckhardt.* Princeton: Princeton UP, 1990.

Glaser, Horst Albert. "Auf Witikos Spuren. Stifters Schlachtenschilderungen." *Germanistische Mitteilungen* 40 (1994): 179–197.

Glaser, Horst Albert. *Die Restauration des Schönen: Stifters "Nachsommer."* Stuttgart: Metzler, 1965.

Görner, Rüdiger. "Granit. Zur Poesie eines Gesteins." *Aurora* 53 (1993): 126–138.

Gould, Stephen Jay. *Time's Arrow, Time's Cycle: Myth and Metaphor in the Discovery of Geological Time.* Cambridge, MA: Harvard UP, 1988.

Graczyk, Annette. *Das literarische Tableau zwischen Kunst und Wissenschaft.* Munich: Fink, 2004.

Grätz, Katharina. *Musealer Historismus: die Gegenwart des Vergangenen bei Stifter, Keller und Raabe.* Heidelberg: Winter, 2006.

Grätz, Katharina. "Musealisierung und Sammlung." In *Stifter-Handbuch. Leben – Werk – Wirkung*, edited by Christian Begemann and Davide Guiriato, 305–309. Stuttgart: Metzler, 2017.

Grenier, Lisa. "Il n'y a que les montagnes qui ne bougent pas." In *Viollet-le-Duc et la montagne*, edited by Pierre A. Frey and Lisa Grenier, 12–21. Grenoble: Grénat, 1993.

Groves, Jason. "Erratische Blöcke: Stifters 'Bunte Steine.'" In *Sema: Wendepunkte der Philologie*, edited by Joachim Harst and Kristina Mendicino, 255–269. Würzburg: Königshausen & Neumann, 2013.

Groves, Jason. *The Geological Unconscious: German Literature and the Mineral Imaginary.* New York: Fordham UP, 2020.

Haag, Saskia. "Versetzt. Restaurierung als Entortung in Stifters *Nachsommer.*" In *Figuren der Übertragung: Adalbert Stifter und das Wissen seiner Zeit*, edited by Michael Gamper and Karl Wagner, 77–87. Zürich: Chronos, 2009.

Haberkorn, Michaela. *Naturhistoriker und Zeitenseher: Geologie und Poesie um 1800: der Kreis um Abraham Gottlob Werner (Goethe, A.v. Humboldt, Novalis, Steffens, G.H. Schubert).* Frankfurt a.M.: Peter Lang, 2004.

Haraway, Donna J. *Staying with the Trouble: Making Kin in the Chthulucene.* Durham: Duke UP, 2016.

Heringman, Noah. *Romantic Rocks, Aesthetic Geology.* Ithaca, NY: Cornell UP, 2004.

Heringman, Noah. *Sciences of Antiquity: Romantic Antiquarianism, Natural History, and Knowledge Work.* Oxford: Oxford UP, 2013.

Horn, Eva. *Trauer schreiben: die Toten im Text der Goethezeit.* Munich: Fink, 1998.

Horn, Eva, and Hannes Bergthaller. *The Anthropocene: Key Issues for the Humanities.* London/New York: Routledge, 2020.

Huse, Norbert. *Denkmalpflege: deutsche Texte aus drei Jahrhunderten.* Munich: Beck, 1984.

Hussain, Nadeem J. Z. "The Role of Life in the Genealogy." In *Nietzsche's On the Genealogy of Morality: A Critical Guide*, edited by Simon May, 142–169. Cambridge, UK: Cambridge UP, 2011.

Kauffmann, Kai. "Mündliche Rede, schriftliche Urkunde, wörtliches Protokoll. Adalbert Stifters Geschichtsroman *Witiko.*" In *Grenzräume der Schrift*, edited by Achim Geisenhanslüke and Georg Mein, 127–140. Bielefeld: transcript, 2008.

Keller, Thomas. *Die Schrift in Stifters "Nachsommer": Buchstäblichkeit und Bildlichkeit des Romantextes.* Cologne: Böhlau, 1982.

Koschorke, Albrecht. "Bewahren und Überschreiben. Zu Adalbert Stifters Roman *Witiko.*" In *Vergessene Texte*, edited by Aleida Assmann and Michael C. Frank, 139–157. Constance: Universitätsverlag Konstanz, 2004.

Koschorke, Albrecht. "Das buchstabierte Panorama: Zu einer Passage in Stifters Erzählung 'Granit.'" *Vierteljahresschrift des Adalbert-Stifter-Instituts des Landes Oberösterreich* 38, no. 1/2 (1989): 3–13.

Koschorke, Albrecht, and Andreas Ammer. "Der Text ohne Bedeutung oder die Erstarrung der Angst. Zu Stifters letzter Erzählung 'Der fromme Spruch.'" *DVjS* 61 (1987): 676–719.

Koselleck, Reinhart. *Futures Past: On the Semantics of Historical Time.* Translated by Keith Tribe. New York: Columbia UP, 2004.

Kröke1, Fritz. "Stifters Freundschaft mit dem Alpenforscher Friedrich Simony." *Vierteljahresschrift des Adalbert Stifter-Instituts des Landes Oberösterreich*, 1955, 97–117.

Kurmann, Peter. "Viollet-le-Duc und die Vorstellung einer idealen Kathedrale." In *Eugène Emmanuel Viollet-le-Duc. Internationales Kolloquium Stiftung Bibliothek Werner Oechslin. Einsiedeln*, 32–50. Zürich: gta Verlag, 2010.

Laudan, Rachel. *From Mineralogy to Geology: The Foundations of a Science, 1650–1830.* Science and Its Conceptual Foundations. Chicago: U Chicago P, 1987.

Lebensaft, E. "Spaun, Anton von (1790–1849)." In *Österreichisches Biographisches Lexikon 1815–1950*, vol. 13 (Lfg. 59): 5–6. Vienna: Österreichische Akademie der Wissenschaften, 2007.

Lee, Dwight E., and Robert N. Beck. "The Meaning of 'Historicism.'" *The American Historical Review* 59, no. 3 (1954): 568–577.

Leniaud, Jean-Michel. *Viollet-le-Duc, ou, Les délires du système.* Paris: Mengès, 1994.

Lepenies, Wolf. *Das Ende der Naturgeschichte: Wandel kultureller Selbstverständlichkeiten in den Wissenschaften des 18. und 19. Jahrhunderts.* Munich: C. Hanser, 1976.

Lewis, Charlton T., and Charles Short. "Restauro." In *A Latin Dictionary.* Oxford: Clarendon, 1879.

Lipp, Wilfried. "Adalbert Stifter als 'Conservator' (1853 – 1865)." In *Sanfte Sensationen Stifter 2005: Beiträge zum 200. Geburtstag Adalbert Stifters*, edited by Johann Lachinger et al., 185 – 203. Linz: StifterHaus, 2005.
Loyer, François. "Préface." In *L'imaginaire scientifique de Viollet-Le-Duc*, by Laurent Baridon, 7 – 10. Paris: Harmattan, 1996.
Lübbe, Hermann. *Der Fortschritt und das Museum. Über den Grund unseres Vergnügens an historischen Gegenständen.* London: Inst. of Germanic Studies, 1982.
Matz, Wolfgang. *Adalbert Stifter oder diese fürchterliche Wendung der Dinge: Biographie.* New ed. Göttingen: Wallstein, 2016.
Maurer, Kathrin. "Adalbert Stifter's Poetics of Collecting: Representing the Past against the Grand Narrative of Academic Historicism." *Modern Austrian Literature* 40, no. 1 (2007): 1–17.
Maurer, Kathrin. "Close-Ups of History: Photographic Description in the Works of Jacob Burckhardt and Adalbert Stifter." *Monatshefte* 97, no. 1 (2005): 63 – 77.
Maurer, Kathrin. *Visualizing the Past: The Power of the Image in German Historicism.* Berlin/ Boston: De Gruyter, 2013.
McPhee, John. *Basin and Range.* New York: Farrar, Straus, Giroux, 1981.
Meyer-Sickendiek, Burkhard. "Intertextualität." In *Stifter-Handbuch. Leben – Werk – Wirkung*, edited by Christian Begemann and Davide Guiriato, 206 – 209. Stuttgart: Metzler, 2017.
Michler, Werner. "Vulkanische Idyllen: Die Fortschreibung der Revolution mit den Mitteln der Naturwissenschaft bei Moritz Hartmann und Adalbert Stifter." In *Bewegung im Reich der Immobilität: Revolutionen in der Habsburgermonarchie 1848 – 1849; literarisch-publizistische Auseinandersetzungen*, edited by Hubert Lengauer and Primus-Heinz Kucher, 472 – 495. Vienna: Böhlau, 2001.
Middleton, Robin. "Viollet-Le-Duc and the Rational Gothic Tradition." Dissertation, Cambridge, 1958.
Middleton, Robin. "Viollet-Le-Duc in the Twentieth Century – a French and Anglo-Saxon Survey." In *Eugène Emmanuel Viollet-Le-Duc. Internationales Kolloquium Stiftung Bibliothek Werner Oechslin. Einsiedeln*, 228 – 251. Zürich: gta Verlag, 2010.
Minden, Michael. *The German Bildungsroman: Incest and Inheritance.* Cambridge: Cambridge UP, 1997.
Minden, Michael. "Stifter and the Postmodern Sublime." In *History, Text, Value: Essays on Adalbert Stifter Londoner Symposium 2003*, edited by Michael Minden, Martin Swales, and Godela Weiss-Sussex, 9 – 21. Linz: StifterHaus, 2006.
Möseneder, Karl. "Schriften zur Kunst." In *Stifter-Handbuch. Leben – Werk – Wirkung*, edited by Christian Begemann and Davide Guiriato, 161 – 166. Stuttgart: Metzler, 2017.
Möseneder, Karl. "Stimmung und Erdleben. Adalbert Stifters Ikonologie der Landschaftsmalerei." In *Adalbert Stifter: Dichter und Maler, Denkmalpfleger und Schulmann: neue Zugänge zu seinem Werk*, edited by Hartmut Laufhütte and Karl Möseneder, 18 – 57. Tübingen: Niemeyer, 1996.
Müller, Susanne. *Die Welt des Baedeker: Eine Medienkulturgeschichte des Reiseführers 1830 – 1945.* Frankfurt a. M.: Campus, 2012.
Murphy, Kevin. *Memory and Modernity: Viollet-Le-Duc at Vézelay.* University Park, PA: Penn State U. Press, 2000.
Nayrolles, Jean. "Sciences naturelles et archéologie médiévale au XIXe siècle." In *L'architecture, les sciences et la culture de l'histoire au XIXe siècle*, edited by Centre

d'études foréziennes (Saint-Etienne) and Ecole nationale supérieure d'architecture (Saint-Etienne), 25–49. Saint-Etienne: P.U. de Saint-Etienne, 2001.
O'Connor, Ralph. *The Earth on Show: Fossils and the Poetics of Popular Science, 1802–1856.* Chicago: U Chicago P, 2007.
Oettermann, Stephan. *The Panorama: History of a Mass Medium.* Translated by Deborah Lucas Schneider. New York: Zone Books, 1997.
Osterwalder, Sonja. "Trostlose Wissenschaft, schreckliches Leben. Adalbert Stifter und Edgar Allan Poe." In *Figuren der Übertragung: Adalbert Stifter und das Wissen seiner Zeit*, edited by Michael Gamper and Karl Wagner, 273–288. Zürich: Chronos, 2009.
Parrika, Jussi. *The Anthrobscene.* Minneapolis: U of Minnesota P, 2014.
Pausch, Oskar. "Neue Quellen zur Biographie Friedrich Simonys." *Jahrbuch des Adalbert-Stifter-Institutes des Landes Oberösterreich* 4 (1997): 94–121.
Peters, John Durham. "Space, Time, and Communications Theory." *Canadian Journal of Communication* 28 (2003): 397–411.
Phillips, Alexander. "Adalbert Stifter's Alternative Anthropocene: Reimagining Social Nature in 'Brigitta' and 'Abdias.'" In *German Ecocriticism in the Anthropocene*, edited by Heather I. Sullivan and Caroline Schaumann, 65–85. New York: Palgrave Macmillan, 2017.
Pichler, Franz. "Andreas Baumgartner und sein Werk zur Naturlehre." In *Stifter und Stifterforschung im 21. Jahrhundert: Biographie, Wissenschaft, Poetik*, edited by Alfred Doppler, Johannes John, Johann Lachinger, and Hartmut Laufhütte, 117–125. Tübingen: Max Niemeyer, 2007.
Plumpe, Gerhard. *Theorie des bürgerlichen Realismus.* Stuttgart: Reclam, 1985.
Porter, Roy. *The Making of Geology: Earth Science in Britain, 1660–1815.* Cambridge, UK: Cambridge UP, 1977.
Powers, Elizabeth. "The Sublime, 'Über den Granit,' and the Prehistory of Goethe's Science." *Goethe Yearbook* 15 (2008): 35–56.
Ragg-Kirkby, Helena. *Adalbert Stifter's Late Prose: The Mania for Moderation.* Rochester, NY: Camden House, 2000.
Rheinberger, Hans-Jörg. *Toward a History of Epistemic Things: Synthesizing Proteins in the Test Tube.* Stanford: Stanford UP, 1997.
Rossi, Paolo. *The Dark Abyss of Time: The History of the Earth and the History of Nations from Hooke to Vico.* Translated by Lydia G Cochrane. Chicago: U Chicago P, 1993.
Rudwick, Martin. *Bursting the Limits of Time: The Reconstruction of Geohistory in the Age of Revolution.* Chicago: U Chicago P, 2005.
Rudwick, Martin. *Georges Cuvier, Fossil Bones, and Geological Catastrophes.* Chicago: U Chicago P, 1997.
Rudwick, Martin. *Scenes from Deep Time: Early Pictorial Representations of the Prehistoric World.* Chicago: U Chicago P, 1992.
Rudwick, Martin. *Worlds Before Adam: The Reconstruction of Geohistory in the Age of Reform.* Chicago: U Chicago P, 2008.
Rüsen, Jörn. "Die vier Typen historischen Erzählens." In *Formen der Geschichtsschreibung*, edited by Reinhart Koselleck, 514–605. Munich: dtv, 1982.
Rykwert, Joseph. *On Adam's House in Paradise: The Idea of the Primitive Hut in Architectural History.* Greenwich, CT: Museum of Modern Art, 1972.

Schepers, Wolfgang. “Zu den Anfängen des Stilpluralismus im Landschaftsgarten und dessen theoretischer Begründung in Deutschland.” In *Geschichte ist allein zeitgemäss*, edited by Michael Brix and Monika Steinhauser, 73 – 93. Lahn-Giessen: Anabas, 1978.

Schlaffer, Hannelore, and Heinz Schlaffer. *Studien zum ästhetischen Historismus.* Frankfurt a. M.: Suhrkamp, 1975.

Schlott, Michael. “Historismus.” In *Reallexikon der deutschen Literaturwissenschaft*, edited by Harald Fricke, Klaus Frubmüller, Jan-Dirk Müller, and Klaus Weimar, 2:58 – 62. Berlin: De Gruyter, 1997.

Schneer, Cecil. “The Rise of Historical Geology in the Seventeenth Century.” *Isis* 45, no. 3 (September 1954): 256 – 268.

Schneider, Sabine. “Kulturerosionen. Stifters prekäre geologische Übertragungen.” In *Figuren der Übertragung: Adalbert Stifter und das Wissen seiner Zeit*, edited by Gamper, Michael and Wagner, Karl, 249 – 269. Zürich: Chronos, 2009.

Schneider, Sabine, and Barbara Hunfeld, eds. *Die Dinge und die Zeichen: Dimensionen des Realistischen in der Erzählliteratur des 19. Jahrhunderts.* Würzburg: Königshausen & Neumann, 2008.

Schnyder, Peter. “Die Dynamisierung des Statischen. Geologisches Wissen bei Goethe und Stifter.” *Zeitschrift für Germanistik* 19, no. 3 (2009): 540 – 555.

Schnyder, Peter. “Geologie und Mineralogie.” In *Stifter-Handbuch. Leben – Werk – Wirkung*, edited by Christian Begemann and Davide Guiriato, 249 – 253. Stuttgart: Metzler, 2017.

Schnyder, Peter. “Grund-Fragen: Goethes Text ‘Über den Granit’ als ‘Ur-ei’ der Wissensrepräsentation.” In *“Ein Unendliches in Bewegung”. Künste und Wissenschaften im medialen Wechselspiel bei Goethe*, edited by Barbara Naumann, 245 – 263. Bielefeld: Aisthesis, 2012.

Schnyder, Peter. “Schrift – Bild – Sammlung – Karte. Medien geologischen Wissens in Stifters *Nachsommer.*” In *Figuren der Übertragung: Adalbert Stifter und das Wissen seiner Zeit*, edited by Michael Gamper and Karl Wagner, 235 – 248. Zürich: Chronos, 2009.

Schnyder, Peter, ed. *Erdgeschichten. Literatur und Geologie im langen 19. Jahrhundert.* Würzburg: Königshausen & Neumann, 2020.

Schößler, Franziska. “Der Weltreisende Alexander von Humboldt in den österreichischen Bergen: Das naturwissenschaftliche Projekt in Adalbert Stifters ‘Nachsommer.’” In *Ordnung-Raum-Ritual: Adalbert Stifters artifizieller Realismus*, edited by Sabina Becker and Katharina Grätz, 261 – 2185. Heidelberg: Winter, 2007.

Schößler, Franziska. “Rahmen, Hüllen, Kleider und das Phantasma der Durchsichtigkeit. Verschwindende Medien in Stifters *Nachsommer.*” In *Rhetoriken des Verschwindens*, edited by Tina-Karen Pusse, 105 – 119. Würzburg: Königshausen & Neumann, 2008.

Schubenz, Klara. “Botanik/Wald.” In *Stifter-Handbuch. Leben – Werk – Wirkung*, edited by Christian Begemann and Davide Guiriato, 257 – 262. Stuttgart: Metzler, 2017.

Schubenz, Klara. *Der Wald in der Literatur des 19. Jahrhunderts: Geschichte einer romantisch-realistischen Ressource.* Göttingen: Wallstein/Konstanz UP, 2020.

Schulin, Ernst. “Zeitgemäße Historie um 1870. Zu Nietzsche, Burckhardt und zum ‘Historismus.’” *Historische Zeitschrift* 281 (2005): 33 – 58.

Schuller, Marianne. “Das Gewitter findet nicht statt oder die Abdankung der Kunst.” *Poetica* 10, no. 1 (1978): 26 – 52.

Schuster, Jana. “Sanftmütige Gärtner, oder: post-katastrophische Kulturation im frühen Anthropozän. Bio- und Geonarrative bei Adalbert Stifter.” In *Repräsentationsweisen des Anthropozäns in Kultur und Medien*, edited by Gabriele Dürbeck and Jonas Nesselhauf, 29–47. Berlin: Peter Lang, 2018.

Seibt, Ferdinand. “Stifters ‘Witiko’ als konservative Utopie.” In *Deutsche und Tschechen. Beiträge zu Fragen der Nachbarschaft zweier Nationen.*, edited by Adalbert-Stifter-Verein München, 23–29. Munich: Delp, 1971.

Selge, Martin. *Adalbert Stifter: Poesie aus dem Geist der Naturwissenschaft.* Stuttgart: W. Kohlhammer, 1976.

Selge, Martin. “Die Utopie im Geschichtsroman. Wie man Adalbert Stifters ‘Witiko’ lesen kann.” *Der Deutschunterricht* 27, no. 3 (1975): 70–86.

Sieburth, Richard. “Une idéologie du lisible: le phénomène des Physiologies.” *Romantisme* 47 (1985): 39–60.

Simonis, Linda. *Genetisches Prinzip: Zur Struktur der Kulturgeschichte bei Jacob Burckhardt, Georg Lukács, Ernst Robert Curtius und Walter Benjamin.* Tübingen: Max Niemeyer, 1998.

Stiegler, Bernd. *Philologie des Auges: Die photographische Entdeckung der Welt im 19. Jahrhundert.* Munich: Fink, 2001.

Streitfeld, Erwin. “Aus Adalbert Stifters Bibliothek. Nach den Bücher- und Handschriften-Verzeichnissen in den Verlassenschaftsakten von Adalbert und Amalie Stifter.” *Jahrbuch der Raabe-Gesellschaft* 18 (2010): 103–148.

Tang, Chenxi. *The Geographic Imagination of Modernity: Geography, Literature, and Philosophy in German Romanticism.* Stanford: Stanford UP, 2008.

Thaon, Bernard. “Viollet le Duc, pensée scientifique et pensée architecturale.” In *Actes du colloque international Viollet-le-Duc Paris 1980*, edited by Pierre-Marie Auzas, 131–142. Paris: Nouvelles Editions Latines, 1982.

Thums, Barbara. “Adalbert Stifters ‘Der Nachsommer’: Reste-lose Poetik des Reinen?” In *Was übrig bleibt. Von Resten, Residuen und Relikten*, edited by Barbara Thums and Annette Werberger, 79–97. Berlin: trafo, 2009.

Timpanaro, Sebastiano. *The Genesis of Lachmann's Method.* Translated by Glenn W. Most. Chicago: U Chicago P, 2005.

Van der Steeg, Christian. *Wissenskunst: Adalbert Stifter und Naturforscher auf Weltreise.* Zurich: Chronos, 2011.

Vedder, Ulrike. “Erbschaft und Gabe, Schriften und Plunder. Stifters testamentarische Schreibweise.” In *History, Text, Value: Essays on Adalbert Stifter Londoner Symposium 2003*, edited by Michael Minden, Martin Swales, and Godela Weiss-Sussex, 22–34. Linz: StifterHaus, 2006.

Vogl, Joseph. “Der Text als Schleier: Zu Stifters *Der Nachsommer.*” *Jahrbuch der deutschen Schillergesellschaft* 37 (1993): 298–312.

Vogl, Joseph. “Poetologie des Wissens.” In *Einführung in die Kulturwissenschaft*, edited by Harun Maye and Leander Scholz, 49–71. Munich: Fink, 2011.

Wagner, Lori. “Schick, Schichten, Geschichte: Geological Theory in Stifter's *Bunte Steine.*” *Jahrbuch des Adalbert Stifter Institutes des Landes Oberösterreich* 2 (1995): 17–41.

Wegmann, Nikolaus. *Bücherlabyrinthe: Suchen und Finden im alexandrinischen Zeitalter.* Köln: Böhlau, 2000.

Weigel, Harald. *"Nur was du nie gesehn wird ewig dauern": Carl Lachmann und die Entstehung der wissenschaftlichen Edition.* Freiburg i.B.: Rombach, 1989.
Weigel, Sigrid. *Genea-Logik: Generation, Tradition und Evolution zwischen Kultur- und Naturwissenschaften.* Munich: Fink, 2006.
Weigel, Sigrid. "Zur Dialektik von Geschlecht und Generation um 1800. Stifters 'Narrenburg' als Schauplatz von Umbrüchen im genealogischen Denken." In *Generation. Zur Genealogie des Konzepts – Konzepte von Genealogie.*, edited by Ohad Parnes, Ulrike Vedder, Stefan Willer, and Sigrid Weigel, 109–124. Munich: Fink, 2005.
Wellbery, David. "Die Wahlverwandtschaften." In *Goethes Erzählwerk: Interpretationen*, edited by Paul Michael Lützeler, 291–318. Stuttgart: Reclam, 1985.
Wittkau-Horgby, Annette. *Historismus: zur Geschichte des Begriffs und des Problems.* Göttingen: Vandenhoeck und Ruprecht, 1994.
Zumbusch, Cornelia. "Der Gang der Geschichte: Historismus und genetisches Erzählen in Stifters 'Witiko.'" *Jahrbuch der deutschen Schillergesellschaft* 51 (2007): 227–251.
Zumbusch, Cornelia. "Nachgetragene Ursprünge: Vorgeschichten im Roman (Wieland, Goethe, Stifter)." *Poetica* 43 (2011): 257–300.

Index

https://doi.org/10.1515/9783110685534-009